CALGARY PUBLIC LIBRARY

NOV 2016

## Praise for *Rising Ground*

"Superb."—Robert MacFarlane, *Guardian,* Book of the Year

"Intriguing."—Tom Robbins, *Financial Times,* Book of the Year

"A fascinating study of place and its meaning."—
Justin Cartwright, *Observer,* Book of the Year

"Pitch-perfect prose"—Tom Adair, *Scotsman,* Book of the Year

"Fascinating and hauntingly evocative. . . . A truly wonderful
and enjoyable book."—Jan Morris, *Literary Review*

"Marsden's writing is just so good. Short, pacey chapters
and an intimate and aphoristic style complement his
powerful evocation of different terrains."—*Guardian*

"Marsden writes with charm and passion. . . . Vivid epiphanies
are scattered throughout the text. *Rising Ground* wears its
philosophy lightly and informs as it entertains. It is beautifully
written, and a labour of love."—*Times Literary Supplement*

"A deft and compelling blend of cultural history,
travelogue and observation."—*Daily Telegraph*

"The beauty of Marsden's book is that, although it is thoroughly
researched and rigorously argued, it comes across as the result of
experience, the close frequenting of that characterful region. . . .
The extraordinary richness of daily perceptions and antiquarian
knowledge assembled in *Rising Ground* never feels like a tray of
specimens laid out for inspection."—*London Review of Books*

"It is Marsden's close attention to the immediacy of his experience—the shape of the particular hill, the sound of the curlew's cry in the early hours, the feel of heather crunching beneath his feet—that keeps him, and us, interested in this journey."—*Financial Times*

"A wonderful topographical history of the West Country but also a great essay on why some places exert such a psychological pull on us."—*Wanderlust*

"Every rock, hill and cliff holds a tale or a legend, and *Rising Ground* shows that such landscapes are not just close to our hearts but are a crucial part of our culture too."—*Geographical Magazine*

"Marsden's often solo walk is described in such evocative detail readers can envisage the rough Cornish landscape without previous knowledge of it. His probing yet conversational style is both informative and entertaining."—*Irish Examiner*

"One of those literary wonders that achieves many things. . . . It is, in one sense, a personal, profound book about the nature of belonging and man's desire to lay down roots and explore his surroundings. It is also a book that beautifully and succinctly evokes those surroundings . . . (and often with a refreshing sense of fun)."—*Country and Town House*

# RISING GROUND

*A Search for the Spirit of Place*

Philip Marsden

THE UNIVERSITY OF CHICAGO PRESS
*Chicago*

**PHILIP MARSDEN** is the award-winning author of a number of works of fiction, nonfiction, and travel writing, including *The Levelling Sea*, *The Spirit-Wrestlers*, and *The Bronski House*.

The University of Chicago Press, Chicago 60637
© 2014 by Philip Marsden
Originally published in English by Granta Publications under the title *Rising Ground* copyright © Philip Marsden 2014
First published in Great Britain by Granta Books 2014
All rights reserved. Published 2016.
Printed in the United States of America

25 24 23 22 21 20 19 18 17 16    1 2 3 4 5

ISBN-13: 978-0-226-36609-8 (CLOTH)
ISBN-13: 978-0-226-36612-8 (E-BOOK)
DOI: 10.7208/chicago/9780226366128.001.0001

LIBRARY OF CONGRESS CATALOGING-IN-PUBLICATION DATA

Marsden, Philip, 1961– author.
Rising ground : a search for the spirit of place / Philip Marsden.
pages cm
"First published in Great Britain by Granta Books 2015"
Includes bibliographical references and index.
ISBN 978-0-226-36609-8 (cloth : alk. paper) — ISBN 978-0-226-36612-8 (e-book) 1. Historic sites—England—Cornwall (County) 2. Cornwall (England : County)—Description and travel. 3. Cornwall (England : County)—Antiquities. I. Title.
DA670.C8M34 2016
942.3'7--dc23

2015034535

∞ This paper meets the requirements of ANSI/NISO Z39.48-1992 (PERMANENCE OF PAPER).

*To*
*T. C. M-S.*
*R. F. H.*

*and*

*In memory of D. B. H. 1935–1998*

# CONTENTS

# ILLUSTRATIONS

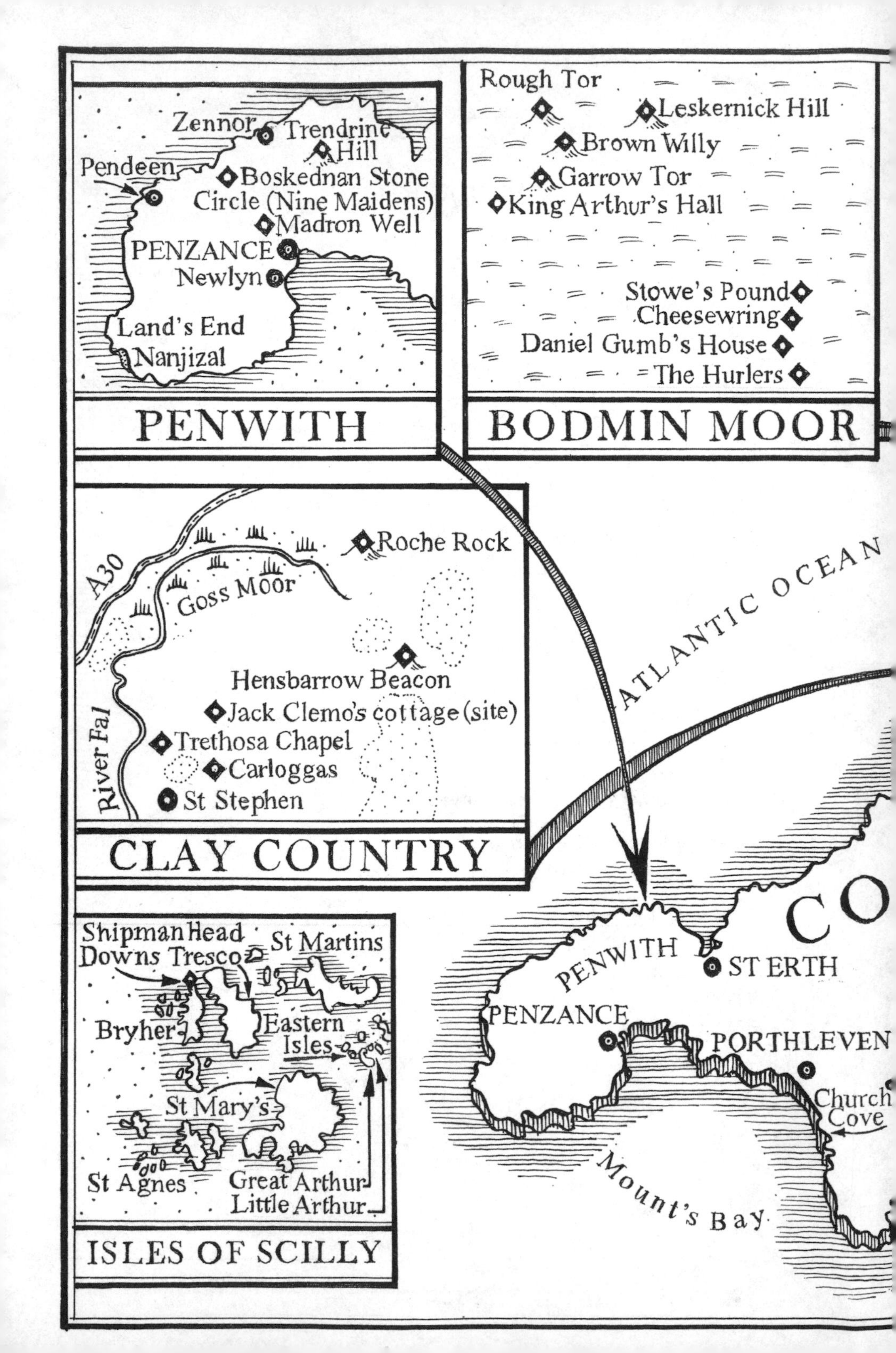

Zennor
Trendrine Hill
Pendeen
Boskednan Stone Circle (Nine Maidens)
Madron Well
PENZANCE
Newlyn
Land's End
Nanjizal
PENWITH
Rough Tor
Leskernick Hill
Brown Willy
Garrow Tor
King Arthur's Hall
Stowe's Pound
Cheesewring
Daniel Gumb's House
The Hurlers
BODMIN MOOR
Roche Rock
A30
Goss Moor
Hensbarrow Beacon
Jack Clemo's cottage (site)
River Fal
Trethosa Chapel
Carloggas
St Stephen
CLAY COUNTRY
ATLANTIC OCEAN
Shipman Head
Downs Tresco
St Martins
Bryher
Eastern Isles
St Mary's
St Agnes
Great Arthur
Little Arthur
ISLES OF SCILLY
CO
PENWITH
ST ERTH
PENZANCE
PORTHLEVEN
Church Cove
Mount's Bay

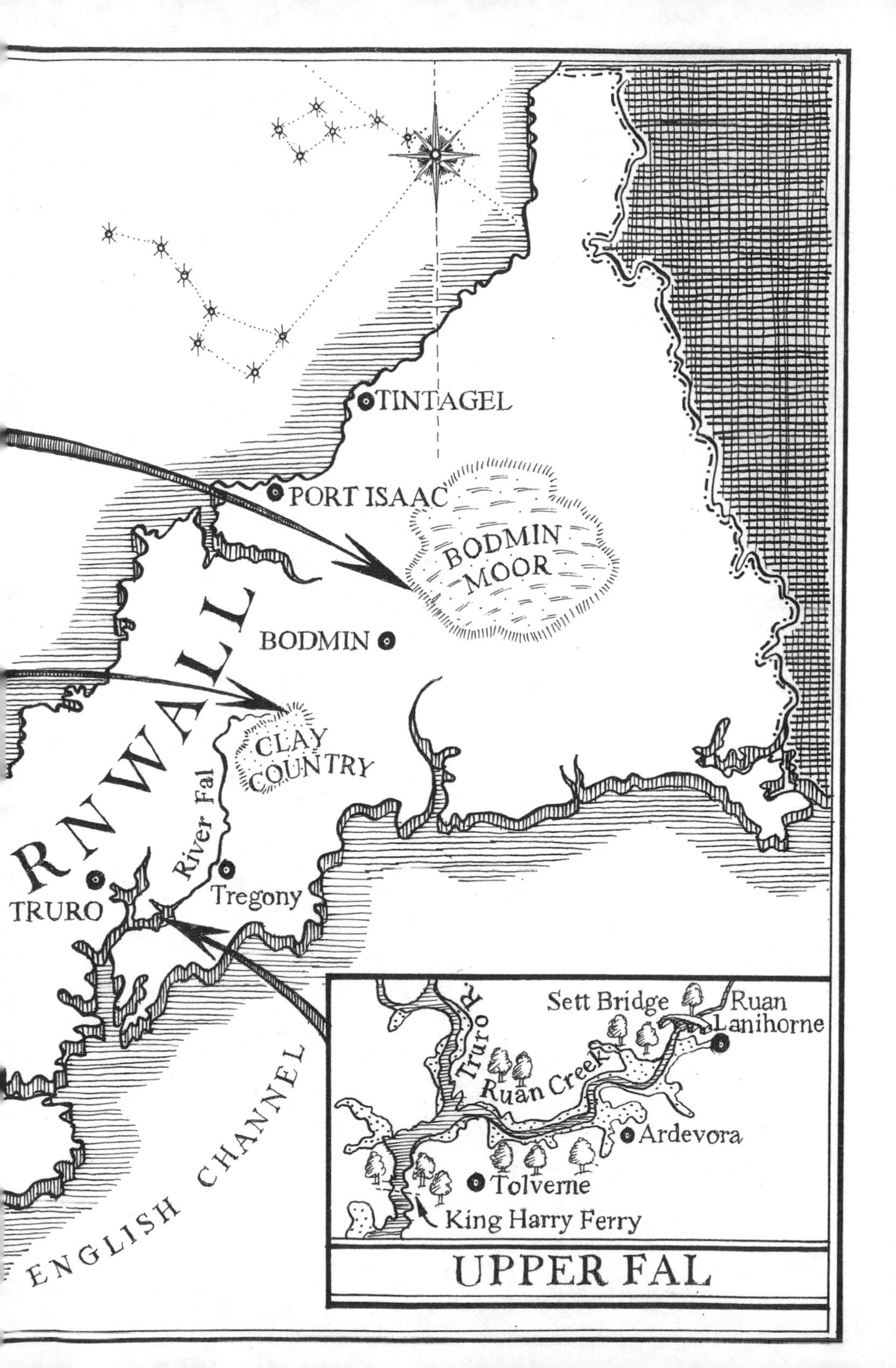
TINTAGEL
PORT ISAAC
BODMIN MOOR
BODMIN
RNWALL
CLAY COUNTRY
River Fal
TRURO
Tregony
ENGLISH CHANNEL
Sett Bridge
Ruan Lanihorne
Truro R.
Ruan Creek
Ardevora
Tolverne
King Harry Ferry
UPPER FAL

# PART I

# 1 | MENDIP

Mendeppe (1185): possibly from Celtic *monith*, 'mountain' or 'hill', and Old English *-yppe*, 'upland' or 'plateau', or from Brythonic *mened*, with an Anglo-Saxon suffix *-hop*, 'valley'. A Basque origin has also been suggested, from *mendi*, 'mountain'.

In the village where I grew up, on the edge of the Mendip Hills, a lane ran up from the church and near the top, it forked. One way curved back into the valley while the other, following the slope, continued upwards. That was the way we always took. The tarmac began to fracture, revealing polished knuckles of bedrock, and then that track too dropped down the hill and a smaller path broke from it, pushing up through the edge of a wood, to emerge – quite suddenly – into open ground.

Years later, I can recall with absolute clarity that moment and the sight which always stopped us dead: the miles of bracken stretching away to the south, rising and falling like sea-swells. No building broke that open arena; no cattle, no sheep and no fences. We never paused for long but hurried on to Long Rock – a dragon-back ridge with its view over the Severn Estuary. Beyond it was a sweep of ground and chest-high ferns that ended with a group of yews. Their roots clawed over the edge of the rock and out into space. We lay on our stomachs and looked down into the gorge, and every now and then, far below, a car would beetle its way along the bottom. The opposite side was scattered with scrub. It was wooded in parts and in others, where it steepened, was covered in rocky scrub and scree. At the top, the land opened out again, and we knew that place as well, with its woods off to one side, and a plantation beyond but everywhere else just open ground; as you climbed the central ride and saw the skyline ahead, wide and flat beneath the clouds, you felt certain that you could carry on that way for ever.

On to these blank places – known only as the 'hill' and the 'combe' – I sketched the map of my childhood. It proved an endlessly supple backdrop to the daydreams of a seven-year-old, or the broodings of a fourteen-year-old. I spent days up there – sometimes with my brother, sometimes with friends, sometimes alone.

It was limestone country, so there were caves. On the hill they were called swallets: sudden pits that looked like outsize basins, dropping to plug-holes that we tried to squeeze into but never dared go far. Down in the gorge was a proper cave called Aveline's Hole. The entrance opened at the base of the cliffs like a giant's mouth. We stood outside, then went in a few steps, then further as the years passed, breathing the stale air, sliding on the buttery mud. The tunnel was neither a walk nor a climb but something in between, a slant down into the gloom. But it went nowhere, tapering like a sock to a damp and rocky end. Lots of things were like that: they seemed exciting to begin with but then they just ran out.

Later when we went potholing, with lamps and helmets and ropes, the ways in were much more furtive. Often they seemed to be in a bush. But they opened up labyrinths of tunnels and chimneys and grottos. Now, with real caves to explore, we walked past Aveline's Hole with hardly a glance. I loved the strangeness of those potholes. They might lead us up and down again by a completely different way, spilling us back into daylight in a different place. Yet in the end the potholing stopped. I think I preferred it out on top, with distance everywhere and the sky.

For years, until I had my own children, whenever I heard the word 'childhood', it always brought to mind a picture of the hill and the combe. I saw the lane running up from the church, and that first sight of the open ground stretching away in all directions. I saw the cliffs of the gorge and the caves, and the hillside beyond them with close-cropped grass and scatters of grey limestone rising towards the sky, and I was always amazed how that simple slope could conjure up so much of a life.

*

In September 2003, a news story caught my eye: 'UK's Oldest Cemetery Identified'.

> A narrow cave in a gorge in Somerset has been identified as the oldest cemetery in Britain . . . Scientific tests, released yesterday, showed it had been sealed and abandoned more than 6,000 years before the first stone of the pyramids of Egypt was laid. The site, Aveline's Hole, is unique in Britain and earlier than anything similar on mainland Europe.

Aveline's Hole! It turned out that a collection of human bones had been found there in the 1920s and placed in the Bristol City Museum. But the museum had been destroyed in a bombing raid in 1941. A few charred remains had been recovered, placed on a shelf and forgotten about. Now using AMS radiocarbon dating, the bones had been taken down and their age identified in a very specific date bracket of astonishing antiquity – between 8460 and 8140 BC.

Shortly afterwards, a paper appeared in the *Proceedings of the University of Bristol Spelaeological Society* – the UBSS – giving a lot more detail. Written by R. J. Schulting, it is a scholarly work of nearly one hundred pages, full of tables and date ranges and graphs of chemical constituents. Schulting has managed to work out not only what the people ate, but where they came from. In the bones' collagen, the proportion of stable carbons and nitrogen isotopes point to a diet in which, oddly, marine food did not feature at all (but the sea was a lot lower then – the Severn Estuary some thirty-seven metres below present levels). Strontium extracted from the enamel of remaining teeth was compared with that of the surrounding geology and a close match found, suggesting that those whose bodies ended up here had lived most of their lives around the Mendips.

But it was less the details of diet or genetic make-up that gripped my attention than the glimpses of possible practice, of motive and belief. The laying of the bodies in the cave was a deliberate act. They were not buried in the soil or left out on rocks for excarnation. They were taken to the combe, into the underground chamber and 'articulated', that is, propped up. A large stone was then used to seal the entrance. Perhaps fifty bodies were taken to Aveline's Hole, even as many as a hundred. It is the earliest evidence in this country of large-scale ritual.

One phrase stayed with me long after reading the report. Aveline's Hole, wrote Schulting, may have been chosen because it was a 'mythic place in the landscape'. Something about the location, about the cave in the combe, had prompted those burials. He mentions the possibility of there being even older human and animal bones in the cave, from the Creswellian period thousands of years earlier.

And from that emerges a dizzying thought – that far back in the ninth millennium BC, the site may have been used because it was already considered old. The astonishment we feel at people performing these rites so long ago might simply be a version of what *they* felt.

Most of what survives from Aveline's Hole is in the UBSS museum and store, hidden away behind the faculty high-rises of Bristol University. One cold February morning, I arranged to meet the museum's curator, Linda Wilson, outside Senate House. As she led me through the alleys she spoke of the Chauvet cave in France – where Werner Herzog made his wonderful film *Cave of Forgotten Dreams* – and how she'd been one of the few invited to visit it and see in the flesh its 40,000-year-old pictures of horses' heads.

We came to a stable block. It looked almost abandoned, with several windows concreted up. Unlocking the door, Linda pressed a switch and the fluorescent tubes flickered, then half-filled the room with light. She looked up at the ceiling, and tutted. 'One of those tubes, gone again.'

A steep wooden staircase led up to a couple of rooms tucked in under the roof. In one were racks and racks of metal shelving and a dusty light falling through the Velux. A pair of microscopes stood on one shelf; on another an old computer monitor, its bulky cathode-ray tube itself a relic from another age. The room was dominated by the shelves and by the archive boxes on them – dozens, marked in black marker pen: *Wookey Hole – rib and vertebral fragments; Totty Pot, GB, Gough's; Rhino and Carnivores, Wolf.*

A stack of ten or more boxes were marked *Aveline's*. Linda stretched to pull the first one down, and lifted the lid. Inside was an assortment of containers – plastic specimen bags, tobacco and pastille tins. A large pale-blue box – *Bristow's of Devon, Assorted Fudge* – had scrawled on it in red felt-tip: ANIMAL FRAGS. The next cardboard case contained human remains – loose teeth, and various bones. Wrapped in old tissue, like a Christmas decoration, was a piece of cranium. It was no more than a saucer-sized fragment. I took it out and placed it in my palm. It domed upwards, away from my skin. Reddish soil was encrusted in its crevices. I looked down at the plate-borders, dark squiggles still visible across the surface, and at the porous-looking bone. It was oddly weightless.

But it was impossible not to think of what it once held – a human brain, that swollen organ that distinguishes our species, whose synaptic patterns held a lifetime of accumulated desires, struggles, frustrations, joys. Among those patterns perhaps was the memory of a funeral procession to Aveline's Hole, the same cave in which the individual would later be placed. The bones' carbon dating suggests a short period of use for the cave, a few generations only, a mini-tradition as enigmatic for witnesses then as it is for us now.

Concluding his article on the finds, R. J. Schulting comes up with a possible 'why' for the choice of site. 'There is something about the early Holocene,' he notes, 'that is rather exceptional.' In the centuries before the cave's use as a cemetery, sea-level rise had been very rapid, squeezing the hunting grounds of what is now the Bristol Channel. Aveline's Hole lay on one of the main routes away from the plains, up into the Mendips and off to the east. Schulting suggests that the appropriation of the cave may have been strategic: one group's placing of their dead in its chamber, sealing the entrance, conveyed the message to others that this place, this route, was 'theirs'.

What struck me most about the theory was that it highlighted the power of place, the accumulated veneration for a hole in the ground. In *An Archaeology of Natural Places*, Richard Bradley examines how certain sites become culturally important simply because of their physical form. The Sami of northern Scandinavia, for example, deposited valued metalwork at *siejddes*, sacrificial sites often 'distinguished from the surrounding landscape by their striking topography', outcrops of rock that might look like people or animals. Caves in Minoan Crete likewise were used for elaborate ceremonies, and in the caves various rocks appeared to attract different sorts of offerings because of their shape.

If Schulting's speculation is right, and use of Aveline's Hole was territorial, then it is the earliest evidence of something that would, over the coming millennia, gather pace. As competition for land grew ever more pressing, each acre of the country was fought over and occupied, claimed and counter-claimed. The system of laws and institutions required to deal with the claims became the basis of the nation. So Aveline's Hole can be seen not just as the first known cemetery in Britain but in a way as the beginning of the island's history – and it centred on a 'mythic place in the landscape'.

The site's significance may well have derived from its past, from the association it had with ancestral use – the human remains from the Creswellian period. But there was something else, something more private, something that – shared between multiple individuals, over generations – produced a tradition: the same awe we still feel at the dramatic features of the land, the combination of wonder and bafflement standing at the entrance of a cave.

I went back to Aveline's Hole that afternoon. Dusk came early; the combe was already filled with darkness. The path to the cavern led down between shadowy patches of leafless scrub. With the hole rising high above me, I stood for a moment looking in. Set against the twilight all around, the blackness inside was of a wholly different grade.

I slipped on a head torch; in its beam the rock glowed pale and yellowy like an old manuscript. I stepped inside. Beneath my feet, I could feel the unevenness of a small gully. I pressed on, down into the tunnel, with a steadying hand on the smoothed-off lumps of bedrock. I remembered the phreatic, throat-like shape from years ago; and the smell – heavy and stagnant – of something I could never quite identify. Half-way to the bottom was a slight recess in the rock. That morning, in the UBSS store, I'd looked at the reproduction of the Reverend Skinner's 1819 sketch of the cave's interior. At this place, he had written '*several skeletons found here*'.

It wasn't the same cave I'd known as a child. The carbon-dated relics now filled its dank vault with story, with a past, and the scale of that past had ramped up its significance. Turning off the torch, I sat for a while in darkness. For a moment, I could see nothing. But then, as if the mind cannot bear too much void, shapes started to appear. I turned and looked up. Silhouetted against the blue-grey disc of night, the entrance-boulder looked like a single tooth.

Outside again, the cold felt sharp in my nostrils. Zipping my coat to the chin, I walked on up the combe. Overhead the sky was clear, a strip of stars corridored by black slopes.

# 2 | ARDEVORA

From Middle Cornish *ar-*, 'before, facing, beside' and *devr-*, water. The name suggests the plural form *devrow*, so 'beside the waters'. In Breton the plural form means 'water-course'. Why the plural was used here is impossible to say; it occurs only in one other Cornish name: Devoran, also a flooded valley in the Fal basin, sharing many of the same features.

We weren't looking to move house. We were perfectly happy living in a Cornish seaside village. Our children had just started at the primary school. We had a little boat, and I thought that after the chaotic years of early parenthood, a degree of control was once again settling over our lives. But that May, Charlotte spotted in the local newspaper an old farmhouse for sale. We arranged to view it – curiosity, nothing more. Yet as we drove down the grass-centred track, and saw the arena of rounded hills and the network of oak-fringed creeks and the first glimpse of the house, its chimneys and slate roof rising from beyond a field of barley, I had the sense that our cosy domestic world was about to be shattered.

I knew the place already, at least from a distance. Ardevora – a silent 'o' and the emphasis on the second syllable: 'ar-*dev*-ra'. There was no road or footpath that came close to it. From a bridge in the woods a mile or two away it was possible to glimpse the fish-coloured roofs amidst the greenery. Once or twice, I'd come up the estuary by boat, and seen the name on the chart. This section of the river was only navigable for a couple of hours each side of high water and if you crept up on a big spring tide, you'd find the oaks with their lower branches underwater. With nothing but the cackle of shelduck, and egrets perched in the trees, it felt like you were pushing into some tropical river system, like the Orinoco or the Congo.

But the tide could catch you out in minutes. Some years ago, I made friends with the skeleton crew of a stranded Russian ship, and when the skipper said it was his birthday, we brought the ship's tender up here at high water and left it at an old quay. From there we walked to the pub. It was late when we returned. In the darkness, the tide was already ebbing fast and took us downstream at quite a lick. Sergei sang a sad Cossack song and his voice filled the creek with the emptiness of the steppe. Then, *bump* – we were aground. We all jumped out – except the skipper, whose job it was to remain sitting in the little boat. We hauled him through the shallowing water, slipping and stumbling, hunting blindly for the channel before it disappeared completely. I remember Sergei singing again when we found the water, and the cold and the mud, and one sole pinprick of light on the wooded shore. That was Ardevora.

The seaside house we were still living in was surrounded on three sides by other houses; the water was on the fourth. In the 1920s a retired naval officer had had the house built, wanting nothing more than to see out the remainder of his days sitting in an armchair, gazing at the sea. The wide windows offered him just that: spectacular views, but they came at a price – winter gales buffeting the glass, whining in the eaves for days on end. Now, up at Ardevora on that first May morning, seeing the farmhouse half-hidden in the landscape, run-down ('unspoilt' in estate-agent speak), I was struck not just by the beauty of its siting but by its pragmatism.

Built at the time before railways made their full impact on Cornwall, the farmhouse was designed for work. The garden was a narrow strip of grass before the proper business of pasture. Mains power only reached the house in the 1980s; its water was still pumped up from a hand-dug well. A field was attached, and it rose slightly – sheltering the house from the worst of the wind – before dropping on three sides to the creek. Standing in the field on our first visit, seeing the house with only the roof and top-floor windows visible, I convinced myself that it represented an ageless integrity with the land around it, and felt sure it would pour beneficence over anyone lucky enough to live there. Such delusions are only possible for the besotted. In the days and weeks that followed, I learnt that 'falling in love with a place' meant exactly that – with all its downsides, its yearnings and mood swings.

Tentatively we put our seaside house on the market, and so began an elongated, one-sided courtship. If I heard mention of the farm's name, or even of the neighbouring village, my heart would jump. Too often my working hours were interrupted by a desk-smothering spread of the 1:25,000 Ordnance Survey. I took imagined walks along the tracks and the woods near the house, or along the creeks. After a year of such suspended passions, the farmhouse was suddenly withdrawn from sale. I felt as if the sky had fallen in. Then it was put back on the market and, fickle as a teenager, I was bubbling with bonhomie again.

Now like a stalker, I began to take real walks down through the woods towards it. I learnt to anticipate the exact point, just under a mile away, where the roof would first appear through the trees (beside the pheasant pens, on the edge of the maize field). The path led down towards a side creek and the house was then lost to view – but I could see the field across the corridor of mud flats and the sessile oaks that bordered it. Every tree and shrub I scrutinised. I knew it was unwise to dwell on something that might never happen – but, well, I couldn't help myself.

Another year passed. Our house did not sell. Viewers came and went. Buyers turned out not to be buyers. The banks froze up. And then, suddenly, it was resolved. A date was fixed. I scrambled to finish the manuscript of a book and sent it off to my publisher just days before the removal lorries arrived. Clearing out years of accumulated junk, burning papers, scooping up yards and yards of books, watching the dismantling of rooms I had known all my life, the stripping of a house that I had once yearned for in the same way, I felt only a reckless excitement about what was ahead. I kept expecting the leg-buckling *coup* of nostalgia, even the tiniest stab of sadness or regret – but it never came.

The morning after moving in, I drove to the agricultural suppliers and stood looking at the tractor bolt-ons and cattle fencing, feeling like a newly elected member of an exclusive club. With only a few acres, such senior equipment was not for me. I went inside to the hand-tool section and picked out a bow-saw, a pruning saw, a Roughneck maul and a pair of Bahco Classic loppers. I bought an Efco chain-saw and a spare chain. Seeing a long-handle sickle labelled *Proudly made in RSA for Africa*, I grabbed that too.

We spent the next few weeks hacking and slashing and felling, slicing through serpentine coils of bramble and ivy, sawing down outsize elder. Single shrubs had merged into thickets. There were weeds everywhere. Vegetable growth had been carrying on with absolutely no regard to the stasis of the housing market. As we cleared, I had the feeling that we were cleaning up some bric-à-brac purchase, watching the emergence of its original form, and sensing in its design the hand of a long-dead stranger.

A porch had been added to one side, but its timbers were rotten. We took it down and reverted to the disused front door, flanked by a pair of high yews. At the other end was a dense copse of holly. When felled, it revealed not only a stash of old bottles, a hoof-pick, a broken trowel, three leather boots, and a rust-gnawed tea urn, but also a double bow of Cornish hedge, haw-topped and flanked with zigzag slates. It offered shelter from the westerly winds and looked like a kitchen garden; Charlotte took stakes and string, and we started measuring out the ground for raised beds.

The house itself had been divided into two at some stage and anomalies remained – an extra staircase, blind doorways, a blocked-off passage. Partition walls had been thrown up, and as we tried to draw up plans for renovation, it was hard in places to see what had once joined up with what. Beyond the kitchen was a warren of little rooms and corridors that baffled us; the original lay-out was only revealed by studying the wear-marks on the slate flagstones.

I crouched down to look at one of these stones. A valley had formed down the middle of it and I laid my hand there, running it from side to side. How many decades of daily use had it taken to form? How many bringings in of milk pails, of sides of meat and muddy vegetables, and how much taking out of slops and bones and peelings? How much toing and froing of farm work and family life? We'd moved to this place with a great fanfare of our own new beginnings, but it had all been going on here for generations.

Long before the farmhouse was built in the mid-nineteenth century, a substantial manor had stood on the site – not exactly here, but eighty metres or so towards the creek. In the diocesan records there remain a few scant references to the house, to its lands stretching many miles to the south, and to the Norman family, the Petits, who owned it all. A strategic position on the river – as well as the ancient Cornish name – suggests long use of the site, and I imagined it as one of those hubs in the nation-of-sorts that once connected estuaries in Wales and Ireland and Brittany, Iberia and Scotland.

In 1420, an application was made by the Petits to build a chapel. But within a century, the estate was breaking up. A generation of daughters married away – the eldest into the Killigrew family, whose lands at the mouth of the Fal were better suited to the new age. The upper reaches of the river, a conduit for Cornish tin since antiquity, were suffering a slow paralysis. Silt was clogging the riverbed, pushing the navigable waters far back towards the open sea.

One evening, working on a length of overgrown wall, I sliced through the stem of a cotoneaster, yanked it out and exposed what looked like part of a large stone basin. I cleared the roots and found it was a piece of black granite, dry-laid on the slate wall. I heaved it free. Upended on the grass, it was clear what it was: a piece of medieval tracery, the top half of a cinquefoil window. The chapel! I ran my fingers along the crescent edges of the rebate. I thought of sunlight falling through the glass, patterning the wooden benches below and morning prayers, and the yards around the building busy with animals and work, and ships at anchor in the deep-water creek, and the mingle of Breton and Cornish, Welsh and Irish.

Knowing a little of the past brought with it the first sense of belonging. In 1954, Martin Heidegger wrote an influential essay called 'Building Dwelling Thinking', in which he explores the close connection between the three '*-ings*' of his title – a connection emphasised by his mannered omission of commas. He takes as his example a two-hundred-year-old farmhouse in the Black Forest. Such a place – with echoes of Ardevora – combined religious belief, domestic life and local topography: 'Here the self-sufficiency of the power to let earth and heaven, divinities and mortals, enter in *simple oneness* into things, ordered the house. It placed the farm on the wind-sheltered mountain slope looking south, among the meadows close to the spring.'

'Dwelling' for Heidegger meant much more than just living in a house. It described a way of being in the world. In Old English and High German, he shows how the word *buan* – meaning both 'building' and 'to dwell' – is linked to the verb 'to be'. (The same is true of Cornish and other Brittonic languages: *bos* in Cornish is a verbal noun meaning both 'to be' and a 'building' or 'dwelling'.) So to be is 'to be *in a place*'. Only by knowing our surroundings, being aware of topography and the past, can we live what Heidegger deems an 'authentic' existence. Heidegger is pretty severe about what constitutes authenticity, but his 'dwelling' does highlight something we've lost in our hyper-connected world, something that I found myself rediscovering that spring down at the end of the long track: the ability to immerse ourselves in one place. Heidegger also wrote: '*Only if we are capable of dwelling, only then can we build*' (his italics). I felt he was pointing his magisterial finger directly at me. I found myself straightening my back. Were we even *entitled* to renovate our house?

We had many visitors those early weeks. They'd heard us going on about the place for so long, now they wanted to see it for themselves. Some, I could tell, were faintly repelled when they arrived – by the isolation, by the dustiness and age, and by the chassis-ripping bumps of the track; a few simply abandoned their cars and walked. But most understood our enthusiasm, or at least were polite to us, and there were a great many tours and suggestions and shared experiences, meals outside and evening fires by the creek.

An odd thing kept happening. I'd be dragging brush to the bonfire, or filling the chain-saw with lube, or lashing at a sea of nettles, when suddenly the memory of a figure or scene from years before would flash up: a woman in a village in northern Syria serving *mezes* on the floor of her home; a night spent in a tiny flat in a Moldovan high-rise with a wolf-like dog chained to the kitchen wall; an elderly Georgian poet quoting Akhmatova in a Caucasian orchard; a street corner in downtown Beirut; boat-builders on the banks of the Nile; early coffee beneath an orange tree in Colombia. I'd spent years swallowing up miles, spending time in one place only long enough to write it up. Now all that wandering was being stirred up by a piece of land you could walk across in ten minutes. Was it arbitrary? Was it some time-release mechanism – like the thirty-year rule for classified papers – that made that period now due for re-examination?

No. I now saw a common thread in those journeys and the books they had produced: the effect that physical surroundings have on individuals and whole communities, the capacity of places to create mythologies around them. Sometimes it was direct – the way that the sea draws out traits in those who live close to it, or the religious diversity that has survived amidst the extreme topography of the Caucasus. At other times the story has lain less in the immediate landscape than in the persistence of a lost one – the longing of exiles for the homelands of Poland or Armenia. The years that I spent writing about and travelling in Ethiopia began, above all, with the realisation that, more than almost any country on earth, it had been painted in layer upon layer of meaning by people – from Herodotus to Coleridge, and from Samuel Johnson to Marcus Garvey – who had never been anywhere near it.

In June came a spell of hot days and clear skies. At night I lay in the heat listening to the sound of the creek birds – to the oyster-catchers and their busy, urgent squeaking; to the pre-dawn chatter of Canada geese as they returned from their night-feeding, like young blades back from town; and the workman-like squawk of the heron and the sevenfold *peep* of the whimbrel. But among them, the curlew's cry was king, strong and evocative and exuberant. 'Mazed as a curly' goes the Cornish expression – 'mad as a curlew' – but I loved it for the way it reached out into the emptiness, for its hermit-loneliness.

On one of those nights, unable to sleep, I rose and went into the field. In the grey light of the moon, the contours of the land enclosed the space more than by day. Across the creek, the woods rose in the darkness like a perimeter wall. I recalled an evening nearly twenty years ago in the Armenian mountains. I was talking to a friend's grandmother, a survivor of the massacres of 1915. She'd been six when she was forced to flee her village near Lake Van. She never saw it again, nor her father. Two uncles were killed by *zaptieh* on the road and her younger sister died from disease. But I realised as she spoke – and it shocked me at first – that the real wound for her was not the loss of her family but of her land. She had recounted the deaths with a muted detachment. But recalling Lake Van filled her eyes with tears; the bird's claw fingers of her hands clenched at her knee: 'I can see it now – blue water and white mountains all around . . .'

I walked out to the brow of the field. In the darkness the limbs of the creeks stretched out in several directions. There was a silence of such intensity that it felt as if the earth itself was standing still. Then came the curlew: a few throaty notes, rising, growing faster. A pause and three more notes. They echoed off the shadowy slopes, and in that echo was the strange sense that the whole place was contained.

I had commandeered a bedroom upstairs as an office. Building quotes, planning letters, renewable energy proposals were piling up impressively on the bed; the bare boards below were dotted with helter-skelters of books. With the manuscript gone, my reading was free for the first time in months to follow its own meandering path. But the waywardness, as waywardness usually does, proved much more revealing than anything planned. The deep undercurrents stirred by the move here led me, via archaeology and natural history, via the tree books of Colin Tudge and Oliver Rackham, the bird books of David Cabot and A. J. Prater, to the notion of 'place'.

Some years ago, in the pages of academic books and journals there was a good deal of discussion about the difference between 'space' and 'place'. Very loosely (and definitions form a large part of such scholarly argy-bargy), 'place' is somewhere distinctive, where people react to and live with the particular topography around them. 'Space' on the other hand is an idealised location, abstracted from the real world, a template which can be dropped over any point on the earth's surface and allow meaningful discourse about it. Most of the recent work on the subject is driven by the conviction that 'place' has been having a hard time of it for too long, and that 'space' should now move over.

The political geographer Arturo Escobar was not alone in finding that the imbalance could be traced far back into the history of ideas: 'Since Plato, western philosophy – often times with the help of theology and physics – has enshrined space as the absolute, unlimited and universal, while banning place to the realm of the particular, the limited, the local and the bound.' The long-term emphasis on 'space' has had unforeseen consequences: monoculture in farming, homogeneous housing, duplicated shopping malls, bio-depletion and the catastrophic destruction of habitats – the abiding sameness that characterises contemporary life. 'The making of standardised landscapes,' wrote Edward Relph in his 1976 *Place and Placelessness*, 'results from insensitivity to the significance of place.'

What appealed to me as I settled into the orbit of my new life was that, for many of these professors of social science, the idea of 'place' appeared to rise above the utility-carpeted corridors of their faculties and offer instruction in the very practice of living. 'To be human,' wrote Tim Cresswell of the University of Wales, 'is to be "in place".' Edward Casey is Professor of Philosophy at the State University of New York, the author of several books with titles like *The Fate of Place* and *Getting Back into Place*, and he wrote: 'To live is to live locally and know first of all the place one is in.'

But the chief cheerleader for 'place' remains the American geographer Yi-fu Tuan. His work is bracing, lively and far-reaching both in its ideas and its sources (Winnie-the-Pooh and Chaucer cited in a single paragraph). One of his books won me over straight away with its title: *Topophilia* ('love of place'). Tuan was not the first to use the term – W. H. Auden coined it in 1948 to describe John Betjeman's particular delight in places. Topophilia, wrote Auden, 'differs from the farmer's love of his home soil...in that it is not possessive or limited to any one locality'. Tuan himself confesses a topophile's love for the desert (purity and timelessness) over the rainforest (decay and death). But his take on topophilia is broader than Auden's. He believes it goes beyond the ability to jet off and wallow in the landscape of your choice, and does originate in the love of home soil. 'Without exception,' Tuan asserts, 'humans grow attached to their native places.'

One of Tuan's distinctions is between two different ways of seeing the world – the 'vertical' and the 'horizontal'. For the bulk of mankind's existence, most people knew only an area roughly equivalent to the distance they could walk in a day; they might be familiar with the route to market or with the landmarks on a seasonal round of grazing – but that was about it. Coupled with a polytheistic belief system – of deities inhabiting both the landscape around them and the heavens above – such groups had 'vertical' perception. They moved, Tuan wrote in a lovely phrase, from 'the shade of the baobab to the magic circle under heaven'. In such animist societies, places are coloured by the gods who inhabit them, and gods take the form of places.

From around 1500 – in Europe at least – a shift began. 'The medieval conception of a vertical cosmos,' Tuan suggests, 'yielded slowly to a new and increasingly secular way of representing the world.' The Americas entered popular consciousness; sea routes to the Far East, Africa and Australia widened it further. Chart-makers' lines spread out across the globe and details of distant places began to populate collective thought. The world altered its plane to the horizontal.

Tuan is giving us a spin on a familiar moment: the great change effected by the Renaissance and the rise of humanism. But it made me wonder again about the role of place: might it be possible to extend the timeline, to assemble a topophile's history? What would a chronology of place look like – of how attitudes towards it have changed over the centuries, and how they've stayed the same? I thought of Aveline's Hole in the long-ago Mesolithic, and then, randomly, of Crusaders crossing Europe to liberate a city they'd never seen, and eighteenth-century antiquarians and crag-happy Romantics, and all those questing souls now re-examining their relationship with the land and the natural world. I thought of the granite cross erected high on top of nearby Dodman Point, facing the ocean with the words: *In the Firm Hope of the Second Coming of Our Lord Jesus Christ*, and of the Petit family, and the household here at Ardevora in the Middle Ages, and what did they feel, and what did they see when they looked out at the ring of hills and the waters of the creek?

For years, I'd been plotting a walk through Cornwall – westward, to the point where the land runs out. Something about the shape of the peninsula invites passage along it: 'stepping westward seemed to be / a kind of "heavenly" destiny', wrote Wordsworth in Scotland. Now I had another purpose: the hunt for a 'mythology of place'. Every topophile knows that some sites are better than others – not just prettier or more dramatic, but endowed with a certain quality that attracts to it a host of stories and ideas. Cornwall has plenty of such places: Tintagel, Land's End, Minions Moor, the mining country, any number of weird and striking rock formations. I'd lost myself often enough in back lanes to know there'd be rich pickings.

Cornwall itself is a good example of the mythologising of place. Few other areas carry such a freight of projections, gathered from so many different periods. A combination of its sea-surrounded shape, its remoteness, its extraordinary range of landscapes and geology, has led to an accretion of myth. Piracy, smuggling, Tudor rebels, Phoenician traders – all these, to outsiders, once coloured its wild shores. As travel grew easier in the nineteenth century, more and more visitors crossed the Tamar to see the place for themselves. For some it was not at all what they'd hoped. Walter de la Mare only felt safe again when back in Devon. Of the far west of

Cornwall, thriller-writer Hammond Innes wrote: 'this is the most god-awful place'.

Others have found Cornwall's shadows more seductive. The novelist Ruth Manning-Sanders was overwhelmed by its 'sense of the primordial, the strange and the savage, the very long ago... something akin to dread'. Jacquetta Hawkes – who made her own poetry from archaeology – felt a deep-down stirring at its very shape: 'Cornwall, a horn of rock... Cornwall is England's horn, its point thrust out into the sea'. Daphne du Maurier was five years old when she heard the name for the first time. Her nurse was telling a visitor that they were going away for the summer – to Cornwall. 'The effect was dramatic, her emphasis on the word Cornwall intense.' The visitor paused, a piece of cake 'midway to her mouth'. '"To Cornwall?" she exclaimed. They exchanged adult glances. Something unspoken seemed to be in the air.'

It was not a good time to take a long walk. I would have to wait until early next year, after the building work. For now, I started to read with more purpose, beginning with the Neolithic. It turned out that the latest interpretation of the monuments – those enigmatic stone circles, stone rows, standing stones and barrows – focused on their siting, on the relationship with other monuments in the area and with natural features. Some three or four thousand years after the ritual use of Aveline's Hole, the idea of 'mythic places' still appeared to be central to collective life. One of the best collections of Neolithic sites in Europe survives on Bodmin Moor: if a sustained walk was out of the question, I could at least slip away every now and then to roam the moor.

No archaeologist has spent more time up on Bodmin Moor than Pete Herring. When I phoned him, it turned out that he lived just a few miles from me. A couple of days later he came over with an off-print from a database he had access to: on it, in the middle of our field, was written '*tumulus*'. It had long since been ploughed in, but it didn't stop us having a look – Pete constantly ducking down low to spot any faint bumps. We found the point on the map, and detected a slight mounding and a possible variation in the grass cover, slightly darker, with a few more germander speedwell.

Pete was drawn to the 'post-processualist' school of archaeology, which aims to understand in human terms the bloodless traces of the past. Most Neolithic sites around Cornwall are now thought to refer visually to prominent hills. Areas like Bodmin Moor and the far west of Cornwall were thick with visual alignments, each appearing to perform some ritual function and converging on a small number of peaks. Many of the cairns outside those areas do something similar. Standing in our field we scanned the skyline, but all we could see that day was summer haze.

Back in the 1980s, Pete had spent three years living alone in a caravan on Bodmin Moor, working on an M.Phil. about Brown Willy's medieval field systems. The caravan was the highest habitation in Cornwall. It was very exposed and one gale-torn night he looked up from his bed and saw the stars – the roof of the caravan was peeling off like the lid from a sardine can. He spent the next few hours chasing the sheets of his work across the moor; for days afterwards he kept finding pieces of paper stuck in the very hedges he was studying.

From there, Pete had built a successful archaeological career, and now travelled the country working in the government's team which assessed the historic landscape – but the place he loved most was Bodmin Moor. He had, after all, spent a very long time up there.

'Too long, probably,' he said with a self-deprecating chuckle. 'By the end I was beginning to lose my social skills.'

A few days later, I drove up to the moor myself.

# 3 | BODMIN MOOR

*Bod*: Cornish, 'dwelling', probably a very early toponym, from *bos*, 'to be'; *-min*: from *meneghy*, 'sanctuary', in this case perhaps less specific than the English suggests, more like 'glebe'. Moor: Old English *mor*, 'waste land, marsh, mountain', in Old Saxon *mor*, 'marsh'. In Cornwall 'moor' also has three other specific uses: (i) a marshy area around a spring; (ii) an area of waste land where tin is found; (iii) the quantity of tin in a lode.

THE DAY WAS WARM, THE sky criss-crossed by streaks of high cloud. Everything else, everything below, was just moor – the smooth slopes, the distant view, the endless stretch of khaki-coloured grass. No fences, no trees, nothing – until the grey uprights of Stannon stone circle. A ewe was bisecting the circle, dragging a bramble that had snagged in the wool of her rump. I counted the stones. It seemed the right thing to do (seventy-six). Some way on, a two-week-old palomino foal stood jelly-legged beside its mother. And then there was lark-song, the wind at my ears, my own footfalls across the tumps, a hidden hollow – squelch! – and faster across the firm ground of King Arthur's Downs.

A novelist friend had given me the name of a woman who farmed in the north-west corner of Bodmin Moor. She had a hut, he'd said, out in the middle of the moor; I was going to ask her if I could use it as a base. I could see her farmhouse a mile or so ahead now, but I was an hour early, so I detoured up to King Arthur's Hall. Some fifty metres long and about twenty wide, the 'hall' was formed by a bank and, on the inside, by a series of inward-sloping stones. The ground enclosed was marshy. No one can agree now what it was used for, nor even what period it comes from – so it has been handed to King Arthur, that doughty old steward of such sites. I lay inside, out of the wind, and chewed on a stalk of sedge and drank in the vastness. Cornwall is a crowded place and its population doubles every summer. Up here it was hard to imagine other people at all. The emptiness was a joy. Bodmin Moor was a place I scarcely knew. I'd always had a bias that pulled me towards the coast, down towards the sea, and boats.

It was afternoon when I walked down to the farm and hobbled over the bars of the cattle grid. In the yard, two dogs uncurled themselves from beneath a sycamore to circle and sniff my boots. Julie had just arrived back. Her cheeks glowed with a day spent beneath the moorland sun. She was rebuilding a mile-long section of stone-faced bank, and she pointed up to the skyline, where a digger made a strange silhouette. Beside it was the peaty scar of her work. She kept cattle, and in the summer they roamed the open spaces of the moor. She was due to go away for a family gathering and gave the impression that any time spent off the moor was under sufferance. The one thing she would willingly leave for was music, to play her cello or to go to a concert.

Julie was happy for me to use the hut and a few days later I brought up food and a bag of clothes and bedding and books and loaded them into the trailer of her quad-bike. She brought a rocking chair into the yard and we heaved that on too.

'The old chair gave out,' she said, throwing a cord over the top of the pile. 'Had to put it on the fire.' She often spent a night or two at the hut, with her books and her music – as if her own moorland farmhouse was not remote enough.

I watched her drive off, and followed on foot. It was late afternoon. The sun was still high over the summit of Alex Tor. The path led over the downs and then dipped to a stream and a narrow bridge. The water ran smooth over a boulder before tumbling into a whisky-coloured pool. Beyond the bridge was a strip plantation of sitka. In the shadows, the ground was soft underfoot and the air syrupy with the smell of resin. A section of bog led out into the sunlight again and I crossed it, hopping from tussock to tussock, before climbing again. Here the slopes were littered with traces of dwellings and ancient fields and I wandered among them. Rabbits scooted off at my approach, like schoolboys caught at some jape.

A couple of days earlier I'd been looking in an archive at the field notes of the tireless archaeologist Dorothy Dudley. In the 1930s she'd spent months up here, excavating the old walls, clearing off the soil to try and recreate the vanished details of domestic life. She later worked with the same diligence all over Cornwall. But it was here, among these now reburied houses, that she chose to have her ashes scattered.

Julie's hut was not actually a hut. It was a small granite house straight from the pages of a folk tale. It looked less built than drawn, freehand, its skewed angles in keeping with the scribbled lines of the moor. A grassy path led to a pair of not-quite-straight columns supporting a porch, with a window on either side and two more above on the first floor.

She'd already dropped off my things and had gone to look for her cattle. The refugee cluster of my stuff stood outside on the grass, around the rocking chair. I felt in my pocket for the key. The door opened on to a ladder staircase in the middle of a single room – cooking at one end, seating at the other. The large granite flags were dark with damp. The east wall was almost entirely taken up with a hearth and cloam-oven beneath a lintel of megalithic scale. Beside it, tucked away, was a delicate recessed window. In its monastic mullions was framed Cornwall's highest point, Brown Willy, from Cornish *bron wennyly*, 'hill of swallows'.

I spent an hour making house – stowing my food, setting up a table near the fire for reading, arranging on top of it the books of moor archaeology, the photocopies and off-prints. I positioned the new rocking chair before the hearth, collected armfuls of gorse stems from the shed, sawed some sitka poles. I filled up a pitcher of water from the outside tap and placed it on the kitchen table – just so, until it looked Vermeer-right, with the late sun on its earthenware curves. Then I climbed the tor.

The house was well sited, snug in the lee of the hill. But as I topped the ridge, the wind struck my face. It hissed in the cavities of bare granite, whose mock architecture suggested ashlar and towers and corridors. I found a slab and perched on its day-warmed surface. Far off, beyond Trevose Head, the sun was dropping towards a bank of pewter cloud. The sea was a strip of gold beneath it. Out there, the beach resorts and the surf hang-outs were in full swing – a world away. Even in July, at dusk, the chill wind cut through my coat.

When I stood to return, the sun had gone; in the gloaming, the moor to the east shrank to a smoky darkness – ageless and without end. Back in the house I lit candles, cooked and chased the damp from the floor with a blaze of gorse, then settled down in front of it to read.

At some point more than six thousand years ago, a strange thing began to happen up here. It all looked different then – valley-bottoms thick with alder, woods of hazel and sessile oak thinning to scrub on the higher ground, and only the ridges and knuckled peaks entirely free of such vegetation. Everywhere the ground was spread with the same lumps of broken granite, snapped from the bedrock during epochs of frost, spread like fallen heroes over the surface of the high moor.

The earliest signs of people here are flint scatters from the Mesolithic era, the chippings of those who followed the seasonal shifts of their own quarry, the deer. By the Neolithic, the visiting groups were following herds of domesticated cattle, moving over the same boulder-strewn ground, starting to clear the trees and open up longer views of the rocky tors.

It is impossible now to know how it began – the practice of shifting those stones, of heaving them up to half-lie on others. Sometimes the aperture that formed in this way allowed the sun to fall through it at a particular hour, on a particular day; or if you sat down you might see it framing a distant tor. Until the last couple of decades no one had ever really noticed these deliberate proppings. One dedicated moor-enthusiast, Tony Blackman, was responsible for learning to distinguish the naturally lying stones from the altered, and for revealing not only the sheer number – hundreds – but also the visual alignments they made with the moorland topography.

In time, the rearrangement of rocks grew more elaborate. Pits were dug, larger menhirs hoisted on their ends, first singly and then in patterns, in rows and circles. Sometimes they were massed into banks that banded the hill-tops. Parties of people dug trenches, erected cairns around small chambers, kerbed them with rocks.

What is baffling to us now is that nothing equivalent in domestic terms survives, that none of this outpouring of effort, the heaving and shoving and hauling, had anything to do with the grind of daily life, with the necessity to eat, to provide food or shelter. The stone arrangements were ritual and Bodmin Moor is now understood to have been a 'sacred' or 'ritual' landscape.

The idea of sacred landscape has only recently slid into the academic mainstream. Until the 1980s, studying Neolithic monuments was all about measurement, chronology and typology: comparison of one stone circle, stone row or chambered cairn with another in some far-off province. But more recently interpretation has become localised, focusing on the monuments' position, what would have been visible from them, how they relate to nearby rivers and ridges and prominent hills.

In Britain there are dozens of examples. At Cranborne Chase in Dorset is a man-made Neolithic causeway or 'cursus' which, when you walk it, reveals the plain chalkland and the numerous barrows in an astonishing and apparently deliberate way. Six miles long, it is the longest cursus in the country. Above Rudston in Yorkshire is a similar collection of cursuses and barrows, laid out, it is thought, to signal the dramatic fluvial features of the Gypsey Race. Stonehenge is set amidst sixteen square miles of ritualised terrain, with dozens of monuments, of which the famous trilithons are only one. Dating suggests use of the site going far back into the Mesolithic, and – in a process common to sacred landscapes – earlier monuments become the focus for later ones. In this way the natural features and the man-made monuments mingle and interact, suggesting that there was little difference in the way they were perceived. Some archaeologists now believe that the entire complex at Stonehenge might be based originally on a glacial fissure which happens to line up with sunset at the winter solstice, and sunrise at the summer solstice.

The spread of relics at such sites is often strictly categorised – weapons deposited in one place, ornaments in another. Materials were likewise divided – wood for the living, stone for the dead. Other parts of the area might be set aside for mortuary use; the burial of men and women was frequently separate.

As more sacred or ritual landscapes are identified around the world, patterns are beginning to emerge, suggesting something universal. Where remnants of traditional belief survive – among the Sami, the Inuit and the Aborigines of Australia – place and the land play a central part in religious life. Historical accounts confirm that many other sites have long since lost the specifics of their sanctity. In the second century AD, Pausanias writes of the mountains of Attica: 'On Pentelicus there is an image of Athena, on Hymettus an image of Hymettian Zeus and there are altars of Showery Zeus and Foreseeing Apollo. On Parnes is a bronze image of Parnethian Zeus, and an altar of sign-giving Zeus . . .'

In a recent study of sacred geography and Hinduism, Diana Eck notes that 'anywhere one goes in India, one finds a living landscape in which mountains, rivers, forests, and villages are elaborately linked to the stories of the gods and heroes'. She records certain patterns, a 'repetition of places, the creation of clusters and circles of sacred places, the articulation of groups of four, five, seven or twelve sites'. Thousands of years of individual visits have embedded reverence: 'This "imagined landscape" has been constituted not by priests and their literature . . . but by countless millions of pilgrims who have generated a powerful sense of land, location and belonging through journeys.'

Neolithic monuments like those on Bodmin Moor are often celebrated as the very beginning of architecture in northern Europe. The crude positioning of one piece of moorstone on top of another is a prototype of the nave at Chartres, the spire at Salisbury or the crescents of Bath. But the 'architecture' does not tell us what preceded it, what led to it, the centuries and centuries of story and song and memory, that poured over the site, and left no trace.

The next day, I left early to walk out to Stowe's Pound. Mist covered everything, wrapping its grey cloth around the shoulders of the hills, filling the valleys – until the sun rose and it vanished. Then there was the A30 to cross at Jamaica Inn and a bit of road-walking, and a couple of dull blocks of conifer plantation. But my memory of that day is not of those things. It is of the endless open, the bare expanses of grassland, a flock of curlew rising near Kilmar Tor, a ruined farm at Rushyford, weed waving in a stream, tormentil in the spongy sward, the warm wind, and that sense of intoxication that comes from too much sun, too much space, or too much time spent on your own – in this case, all three.

I followed no path. I crossed lines of grassy banks, passed low mounds. A cuckoo put its two-note query to the morning. On the edge of Craddock Moor, I stood over a sheep's carcass: the head was gone and the blood-crusted trachea flopped out over the neck. I puzzled at numerous lumps of granite – and began to identify various coffin-size boulders that had not been arranged by gravity.

Once you think you see one propped stone, you see them everywhere. You spot design in the clitter-falls, deliberate curves on the open terrain, shapes in the clouds. Something about the blankness of the moor makes you react against it and fill it with significance – or anything. I remembered a passage in an essay by Jonathan Raban. He was sailing for the first time off the Pacific coast when his depth sounder suddenly showed the water shallowing. The sea bed was rising towards him. A group of rocks? Shoals? Grabbing his chart, he checked his position. Rather than shallowing, he saw that the bottom had in fact dropped off the edge of the continental shelf. Faced with nothing, the sonic pulses were turning the tiniest speck in the water into something significant.

Stowe's Pound is a tor enclosure, one of the earliest types of Neolithic 'building'. A Cornish variant on the 'causewayed enclosures' found both in southern England and in northern Europe, tor enclosures are – as they suggest – loose-stoned walls built around a tor. Likewise the tops of Windmill Hill near Avebury, Hambledon Hill in Dorset or Knap Hill in Wiltshire are all ringed with earthworks. Often the enclosures were rebuilt and reused. During the Bronze Age and Iron Age the walls were reassembled or contained within an outer ring of walls, further muddying questions of form and function. As the first large-scale structures anywhere in the country, the first group effort and intent, you can't help looking at them and imagining that they must say something wonderfully profound. But what exactly? Archaeologists hedge their bets. Tor enclosures, explains one, are 'central places of some kind (or several kinds)'.

I began the short climb up from the moor to Stowe's Hill. Looking up at it, my gaze was absorbed not by the bank itself, which is hardly visible, but by the rocks of the summit, the weird boulder-stacks of relict granite. Their shapes flowed and shifted against the sky. I found it hard to think of them as purely natural, to picture the seeping and freezing and cracking and dissolving that had moulded them. Instead they became animate: a rhino head, a squatting frog, a fat-lipped cod, the helmeted bust of a mythical warrior.

The tallest of them is known as the Devil's Cheesewring, a tottering Jenga of weathered slabs, increasing in size from an impossibly narrow base. It is one of the greatest of Cornwall's gallery of rock curios and most of those who have explored and written about the county in the last few centuries visited it. The roving doctor W. G. Maton wondered in the 1790s about the rocks' provenance, before deciding in the end they were not assembled by man. 'Probably constructed by nature herself, in one of her whimsical moments.' That unflagging Victorian walker and writer George Borrow came here when visiting cousins. (His father was from a village a few miles away, but he'd fled Cornwall for ever after a brawl.) Borrow saw in the Cheesewring an important ancient site, briskly concluding it was man-made: 'evidently . . . the grand place of Cornubian Druid worship'. Wilkie Collins was equally impressed, but thought it natural: 'the wildest and most wondrous of all the wild and wondrous structures in the rock architecture of the scene'. But it made him nervous: 'When you first see the Cheesewring, you instinctively shrink from walking under it.'

I passed through its shadow, and came out on to the hill-top. The moor's slopes dropped to the east, giving a bird's-eye view of the basin of the Tamar and, beyond it, the reclining giant of Dartmoor. From inside the enclosure, you can see the shape of the bank that surrounds it (oddly, the Cheesewring itself is just outside the enclosure). No more than a knee-high remnant of the bank remains – made up of stones of a strikingly uniform size. But it is estimated that it stood over three metres high – a wall for those inside that obscured all but the top of the Cheesewring and the sky.

No excavation has been conducted up here to try and flesh out the story of the enclosure. But framing a hill-top like this can be understood, in broad terms, as a ritual act. Some archaeologists, like Richard Bradley and Chris Tilley, consider that the wall created a sanctum of sorts, offering access to some, denying it to others, implying the existence of a hierarchy. They note, too, a curious pattern. Where tor enclosures occur in Cornwall, they are almost without exception near quoits – those distinctive Neolithic structures made by balancing vast capstones on a stone plinth, creating chambers beneath. Trethevy Quoit is a few miles to the south of Stowe's Pound. The juxtaposition offers the possibility, they believe, that the quoits were built in *imitation* of the tors. They go further. Both Bradley and Tilley suggest that the tors themselves, and the distinctive formations like the Cheesewring, may well have been perceived not as natural forms but as the work of former inhabitants. Until recently folk traditions saw the hand of giants in the tors' construction. Quoits and tor enclosures may thus be seen as a bid to appropriate the power of the land, like those who used Aveline's Hole to deposit their dead.

However compelling, such analysis of specific intent remains speculative. What can be said more broadly is this: the building of tor enclosures was itself speculative. Assembling stones to frame a hill-top is the sort of irrational act people have always performed when dealing with the mysteries of existence. Neolithic man was simply trying to make sense of the world and our place in it.

A woman stepped into the enclosure. We nodded at each other in silence, like visitors in a gallery. Gripping the rim of her hat she turned slowly to take in the view. Then she picked a flattish stone and added it one of the waist-high cairns, the mini-Cheesewrings that had been put up by recent visitors. The stone kept falling off and she bent down close to position it. Very gently she released her finger and thumb. She held them there for a moment. This time the stone was still.

Just below the tor enclosure and the Cheesewring is more evidence of cosmic speculation. In the 1730s a man named Daniel Gumb lived up here with his wife and children in a house he built from boulders. Part cave, part hut, it protected the Gumb family below with a flat and monolithic roof of granite. On cloudless nights, Daniel was said to perch on this roof to calculate the movements of the stars.

Scrambling up on top of it now, I saw chiselled into the stone his own neat carving of the forty-seventh proposition from Euclid's *Elements,* Book I, the so-called Theorem of Pythagoras. Looking down on the moor below, I could see the traces of an earlier geometry – three stone circles arranged in a row, known as the Hurlers.

Daniel Gumb was a figure of some repute in his lifetime. A newspaper report used tabloid language to describe his wife as a 'buxom lass'. Even so, many visitors came up to see him out of curiosity not for his 'love in a cave' but as the 'mountain philosopher'. They were intrigued that this moor-dwelling sage was neither a church-goer nor a dissenter but nourished his soul purely by astronomy and an 'extraordinary love of reading'.

'Daniel,' boasted his wife, 'was far enow better scolard than the passen was.'

I ran my fingers along the stone gullies of Gumb's diagram. He was a mason by trade and it is a bold and accurate rendition of an image familiar from a thousand geometry primers – proof that the square of the two shorter edges of a right-angled triangle added together equal the square of the longer.

I could think of no more powerful image of man's response to the world than the questing Gumb, up here in the years of the English Enlightenment, gazing at the stars and dwelling on the same questions that puzzled the pioneering scientists of his era. It was the same urge that drove our Neolithic ancestors to arrange the moorstone into circles at the Hurlers, to build the wall around the tor – the same questions that tease us now: what law, what force, what patterns exist in the vastness of space? And always behind the questions, the doubt, the depth-sounder beam probing the emptiness for something solid, the fear that there might be none of these things at all.

# 4 | GARROW TOR

Garross Moors (seventeenth century): Middle Cornish *garrow ros*, 'rough hillspur'; *torr*, English, 'rock outcrop'.

Back home, we were preparing for building works. We'd been living in the house as we'd found it, and a part of me wanted to leave it as it was, to let the children carry on drawing dragons on their bedroom walls with felt-tip pens, not to worry about the paintwork, the gaps in the doors or the rot in the porch, to carry on putting wet boots on the disused Lister pump and tying another piece of binder twine to secure the pipe beside it which brought the house's only water supply up from ten metres down the well. But I knew also that the house was slowly falling apart. If we did not do certain things pretty soon, the damage would accelerate.

Arriving in spring had added another layer of exhilaration to the discovery of the house and its surroundings. Blossom, flowering shrubs, rabbits and fallow deer, badgers and stoats, the curlew and the mallard, the shelduck and their trailing crèches of chicks – all of these emerged in ways familiar to them but completely new to us. Nor did nature confine its wonders to the outside. I did nothing about the wisteria shoots that grew through the window of our bedroom, pushing towards the furniture with their slender fingers. A tiny bramble – thorns still pliable, leaves innocent green – had sprouted from a crack in the sitting-room wall, and although a good part of my day was spent cutting back its cousins, this one had a rarity that made me treasure it.

Some intrusions weren't so welcome. 'Dad,' said Arthur one morning. 'There's like a big mouse in the hall.' Now the rustlings and nibblings we heard at night made more sense. I bought some rat-traps. The roots of the exotic shrubs – the cordyline, the fig and fuchsia around the house – were pushing through the stonework and into the void beneath the sprung floor of the parlour. A climbing hydrangea had spread high up the wall of the house, establishing its own little habitat of mosses and wrens' nests. Inside was a corresponding patch of blistered plaster. Up in the roof space, I found not only the hydrangea's tangle of top-growth, but a two-metre mound of dried sticks against the gable: jackdaws had clawed years of nesting material in through a hole in the eaves.

One day, I found a swift on the floor of Clio's room. I placed it on the window-sill, on its short, useless legs. Swifts need at least six metres to launch, and that was about what it had. I watched it plunge to the ground before – just in time – it resumed its high-octane flight. Then there was the hen pheasant we found in the smallest of the upstairs rooms, all flap and cluck at being disturbed, hurrying out through the window. That evening, in the bed-sheets, amid a scattering of dun-coloured down, we found the perfect ovoid of her egg.

It couldn't last. A builder friend pointed out the damage that was being done. When the plumber came to fix a leak, he shook his head at the homemade pipework, the sink that drained straight into an old flower-bed. All the electrics fused one afternoon. Candles added ambience, but having no water did not. The well-pump was electric and, with dishes, bodies and clothes all needing to be washed, the pleasures of a private water supply started to sour. It left a niggling anxiety about the future of the well. ('Always been trouble with that well in dry years,' said one doom-monger who knew the house.) Trying to buy replacement fuse wire proved tricky. I took in the ceramic fuse and the assistant held it up as if it had come from a museum. 'Where on earth did you find that?'

The honeymoon was over. Every wire, every pipe in the house needed replacing. The heating system was not worth sparing. Days were filled now with budgets and estimates and telephone calls about tendering and the technicalities of building work and planning. For the first time in my life, I was aware of the complexity of domestic expectations – the intricate tasks of keeping the weather out and providing heat and water and power. Much simpler was my outside work, the lopping of brambles, the clearing of rotten elder, the sawing and tugging, the barrowing and burning.

Sometimes in the evening, standing before the house, I indulged a certain satisfaction with what we'd done so far – the specimen plants unclogged, the roof-scratching branches removed, the old herring-bone walls revealed. There were moments too when I remembered first seeing the place so surrounded by trees and billows of green that it looked itself like a topographical feature, merged and moulded into the folds of the land. All these plans of intervention, of modifying the house to our own fancies, made me nervous.

One of the more poetic theories for the emergence of Neolithic monument-building is that it was an act of propitiation. The idea hinges on the apparent coincidence of these stone constructions with the introduction of 'settled' agriculture – the clearing of woods, the planting, the keeping of herds. The theory goes that such a shift, from hunter-gathering – which does not alter the land – to practices which depended on clearing and planting, generated a collective anxiety. The ancestral spirits had been disturbed, the power of the earth had been harnessed – and some corresponding gesture of reverence was needed to offset it.

It is a theory that says a good deal more about our own post-industrial guilt than anything likely about the Neolithic age. Yet seeing the house now, stripped bare, and all the tree-stumps and exposed soil, the bonfire mounds of brush and the log piles, I felt a pang. What would the people who'd built it, who'd dwelt here in a state of Heideggerian virtue, make of our washing machines, our planned ceiling lights? And in those moments I could understand a little of the urge to erect a sacred mound in imitation of imagined ancestors.

It was late August before I returned properly to the moor. I'd taken day trips up there – to meet various enthusiasts and seek out their favourite places: abandoned quarries, windy tors and a whole host of visual alignments. Now I'd found a few clear days and, leaving the car at Julie's farm, I walked the familiar path over King Arthur's Downs and across the stream in high spirits.

The door of the house stuck, then, with a shove – opened. There were cobwebs, dead flies, and in the mullioned window a small tortoiseshell beating its wings against the side of Brown Willy. I unpacked my supplies, laid out my sleeping bag and climbed the tor. Evening caught me out – so early, so dark! Back inside I dug out the altar candles I'd brought up, and spiked them on to the wall brackets. I gathered an armful of wood from the old shed and soon the fire was crackling in the grate. Leaning in under the lintel, I watched sparks rise up the chimney-shaft, swimming towards the open air in darting orange shoals, there to be swept away by the wind. I pulled the chair close in to the fire. Outside, the wind rushed at the house, gusting in the eaves as it passed. The logs whispered at my feet, and the chair's rockers – *da-dum, da dum, da-dum* – bumped on the uneven geography of the floor.

I opened a copy of *Stone Worlds*, the account of a dig at Leskernick Hill in the north-east of the moor. For five seasons in the 1990s, a team led by Chris Tilley, Barbara Bender and Sue Hamilton studied the layers of prehistory at Leskernick – the Bronze Age village and surrounding network of Neolithic ritual sites. They combined conventional trowel and trench work with innovative attempts to understand how the landscape might have been seen five thousand years ago. They erected red flags along the stone row and wrapped prominent rocks in coloured cling film. The team was large and human frailties often got in the way of understanding the past. The anthropologists fell out with the archaeologists. The younger workers had their own issues and one of them, Jan Farquharson, wrote a poem: '*The landscape is the meaning, evidence / of how they must have thought and felt,* we say / Now context is all the truth. And it makes sense when cups aren't rinsed, our post-processualist / won't dig, Alan won't wash, Dee's on the fence / And Jane's in love with Una...'

Several times in the night, I woke to the sound of hail at the window above my head. Then at dawn – *bang*! Spread-eagled against the glass, black with the blue glow of day beyond it – a crow.

A few hours later, I set off for Leskernick Hill. The night's gale had eased, but a low cover of cloud still raced overhead. Rays of sunlight pierced its gaps, raking across the grass below like searchlights. A wooden footbridge led over the De Lank river and the path rose towards Brown Willy. I heard sudden breathing at my shoulder, and turned – a jogger. I'd spotted his stick figure, just moments before, it seemed, on the skyline of Garrow Tor.

'Yes, that was me,' he said – and was gone, calling over his shoulder. 'Have a good one!' His was the only voice I heard all day.

Just below Brown Willy, I topped a ridge and could suddenly see far to the north-east, the moor rolling away in wave-like folds, tinged with purple moor grass, towards the distant slopes of Leskernick Hill.

If enclosures like that at Stowe's Pound marked something of a turning-point in our relationship with the land, a beginning of the urge to shape it, then Leskernick is an example of what happened next. For two thousand years, ritual assemblages of rocks were made all over north-western Europe – hill-top enclosures and cairns, then stone rows and cursuses, barrows and chambered tombs, lone standing stones, spectacular sites from Carnac to Maes Howe, from Stonehenge to New Grange. There were megaliths in northern Spain, miniliths on Exmoor, and an untold number of stone circles in Britain, traces of nine hundred of which remain. The Neolithic was less a building boom – two millennia is a long boom – than a process, a gradual elaboration of particular places: monuments built at the site of earlier monuments which had themselves been built to signify the sanctity of natural features.

Amidst the grassy undulations of this corner of the moor, the south-west slope of Leskernick Hill stands out as a grey scar of scattered rocks. The stoniness is extraordinary. Thousands of boulders are caught and frozen in a mass rock-tumble, up-ended, skewed, bunched up in lichened floes of granite. It's not an easy place to walk around – every surface shifts and chinks beneath your weight. But here and there are signs of order. I had my copy of *Stone Worlds* and, using its sketch maps, managed to pick out the outline of huts and compound walls.

Stand here and swivel round through three hundred and sixty degrees and you begin to understand why the site was chosen. Leskernick Hill is in the middle of a ring of higher hills, which create an impression of enclosing the site. 'A nested landscape', is how it's described in *Stone Worlds*. Elsewhere in the book is another image: 'Leskernick is the *omphalos* of the saucer, the Beacon, Tolborough Tor, Catshole Tor, Brown Willy, High Moor, Buttern Hill, Bray Down and Carne Down form the rim.' On the skyline are also a total of twenty-one large cairns. It's a strange feeling, knowing that they're there, an awareness not just of conferred belonging, but of scrutiny – like the gaze of high-hung portraits in an ancestral home. The longer you stand there the more it builds – the awareness of being inside a natural arena; and as it builds, you find yourself scrabbling around for meaning, for some sort of deliberate pattern to explain it.

Fifty years in a similarly contained landscape in Herefordshire prompted Alfred Watkins's sudden revelation that the land was criss-crossed by ancient alignments, by 'ley-lines'. He begins *The Old Straight Track* by describing the setting: 'The pleasant land which has been my field of work is bordered by a broken ring of heights looking inward over lesser wooded hills and undulations.' In Greece, the sites of ancient sanctuaries were often placed where distinctive hills surrounded them. The architectural historian Vincent Scully wrote of Cretan palaces: 'From roughly 2000 BC onward, a clearly defined pattern of landscape use can be recognised at every palace site.' The recurring features are: first, the palace in an enclosed valley; second, a rounded hill either to north or south of the palace; and lastly a more distant hill or mountain – 'higher, double-peaked or cleft'. Such regularity of siting was not aesthetic but a recognition of sacred forces contained in the features of the land. 'The natural and the man-made create one ritual whole,' concludes Scully, 'in which man's part is defined and directed by the sculptural masses of the land and is subordinate to their rhythms.'

Long before the settlement appeared here at Leskernick, the entire area was sanctified. Scattered around the hill and in the valley below are a stone row, a couple of stone circles, many smaller cairns and a very large propped stone. All these – the monuments, the settlement, thousands of years of reverence for this place – derived ultimately from the simple arrangement of hills.

To understand the current thinking on ritual landscape, the experience of the place is everything – and it begins with visibility. Walking the stone row below the settlement, following it westwards, an odd thing happens. As you reach the end, the top rocks of Rough Tor (the first syllable pronounced 'row', as in 'argument') begin to appear. They are still some three miles off, but they sit there above the smooth ridge, growing as you complete the walk, detaching themselves from the unseen hill below them like a row of slowly raised hats.

A summer of intermittent roaming around the moor, days walking alone and with archaeologists and local experts, had left me attuned to its lines of 'intervisibility'. My reaction to the first few was sceptical, but now I found myself constantly trying to line stones up with the skyline. By sheer weight of numbers the alignments had won me over. From the stone circle of the Trippet Stones, if you look north, the summit of Rough Tor – just the summit – sits on the skyline. Walk eastwards, down to the stream and (more or less) at the point it disappears behind the slopes of Hawk's Tor, is a low standing stone. Continue eastwards for half a mile or so and enter the circle of the Stripple Stones, and at that moment Rough Tor reappears, resting its crown on the opposite slope of Hawk's Tor. At Hawk's Tor and Alex Tor are similar 'view-frames' of Rough Tor. On Brown Willy there's another. On a ridge above Withey Brook Marsh is a shifted, head-high rock that marks the exact place where the northern peak of Rough Tor comes into view. Far to the south of Rough Tor is the Cannon Stone, positioned in such a way that, when I wriggled beneath it, I found myself sitting on a natural bench, and, like a projectionist before a flickering screen, could see the summit of Rough Tor, perfectly framed, seven miles to the north. Another stone row rises to the peak of Tolborough Tor, where, several miles away, the top couple of metres of Rough Tor clear the ridge of Brown Willy.

Just above Leskernick is a spectacular 'propped stone', a vast boulder levered up and kept there by the insertion of three smaller rocks at one end. No natural process could have created such an arrangement; it's possible that it predates everything in the vicinity. The very top of Rough Tor is just visible from it, while many of the later monuments are placed precisely where the propped stone itself becomes visible.

No one knew about the Leskernick propped stone until one spring day in April 1995, when a group of walkers rested there. The group was led by Pete Herring and Tony Blackman, who between them probably combined more knowledge of the moor's visual alignments than anyone since – well, since prehistory. They could see the propping of the stone was deliberate but they each said, 'Modern,' attributing it to the Royal Engineers who had been using the area for exercises. But later that day, they noticed that various barrows referenced it. At the summer solstice that year, Pete revisited Leskernick and found that, as the sun set, it shone through the aperture in the propped stone in a way that it wouldn't on any other day. But it wasn't an exact fit. So he contacted the Nautical Almanac Office at Greenwich. Using a theodolite, sixteen-figure grid references and measurements on the stone to the closest millimetre, he sent the information to Greenwich. They calculated that with the shifting of the earth's tilt, the centre of the sun would have pierced the aperture on the evening of 21 June 2851 BC – bang in the middle of the Neolithic.

Over the course of the five-summer study at Leskernick, the propped stone became something of an emblem for the various teams. Its presence hovered over their fieldwork, their comings and goings around the settlement, just as it would have done for people five thousand years ago. Their reaction was to wrap it in cling film and, like a piece of Neolithic bling, spray it gold.

It was after six when I left Leskernick. The low sun flashed on spears of grass. I looked north and watched a lone rider gallop up the long bare slope beneath Brown Willy. His dog detached a steer from a herd of cattle and I saw the small quick dot of the dog drive the slow dot of the cow back towards the rider, and then they all converged and trotted off to the north-east, each following its own stretched-out shadow into the huge evening.

Strange how deserted the moor now is, given the activity of the distant past. The higher areas had been abandoned by the post-Roman centuries, as the climate cooled and people moved to the lowlands. With them went any residual awareness of the monuments. When, briefly, settlement took place in the Middle Ages, farmers brought a different god and a set of sacred places far off in the Holy Land. The ancient monuments were ignored. Why now, after so long, do we have eyes to see the alignments and the sacred landscapes?

The next morning, I left Julie's house and walked around the southern slope of Garrow Tor. The sky was clear, and made shiny blue discs of each of the bog puddles. I was due to meet one of the moor's most committed researchers – Roger Farnworth, a retired schoolmaster. When I'd visited him first at his home in Warleggan a couple of months ago, we'd talked solidly for three and a half hours. We only stopped when he suddenly jumped up and announced he was late for a meeting. We'd talked on the phone since, and walked to various sites and talked, and now he wanted to show me Garrow. Roger was not a trained archaeologist but his diligent reading and years of exploring had led to numerous discoveries. His theories were now attracting the attention of academics.

I waited for him at the stone circles near King Arthur's Downs – 'circles' is perhaps overstating it, they're little more than a few cattle-shunted stumps. I spotted Roger in the distance – a compact figure in a tweed hat amidst the plain expanse of the moor. He started speaking as soon as we were in earshot, and together we crossed the stream and climbed through Garrow's Bronze Age hut circles and field boundaries. Roger was putting forward his views on how a site became sanctified. 'A place achieves prominence for whatever reason, maybe a natural formation. They build a monument to honour that, then the monument adds to the prominence. Like a peacock tail.'

'What?'

'Peacock tail – theory of runaway selection?' He cocked his head at me, then explained: extravagant plumage in male birds has been shown to develop exponentially, where genes for large tails, and for females' preference for large tails, are passed on together. The tail goes on developing until it meets the body's constraints for carrying such flamboyance. Likewise rock arrangements in the Neolithic, growing more elaborate over time, if not as brightly coloured.

The top of Garrow is a long grassy plateau flanked by wind-rounded and lenticular outcrops of granite. They look strikingly architectural. A few weeks ago, I'd come up here with my family. Charlotte and I had had to use every trick of bribery and coercion to persuade the children up the lower slopes. But as soon as they saw the rocks on top, they ran towards them: 'A palace! Look – a throne, towers, dungeons!' (Something similar has happened at other 'ritualised' natural sites – the Logan Rock near Treen, Rough Tor and the fogou at Carn Euny.)

To me, the top of Garrow was like the aisle of a great open-air cathedral, culminating in a chancel slab of bed-rock entered between two piers of granite. As you pass the piers, it feels like crossing a threshold. In front, with the land dropping away, is a view of Rough Tor so perfect, so deliberate-looking, it is as if the entire end of the hill has been placed there simply to present the view to maximum effect.

Roger was convinced that Garrow represents one of the most important ceremonial sites on the moor. He'd brought up a geologist in order to identify those rocks that had moved through natural processes and those that had been 'arranged'. Now Roger led me off the peak to a hidden cave he'd found, partly the result of someone sliding a vast wafer of granite in front of it. Like everything else here, the slender opening of the cave faced Rough Tor.

We sat in the lee of the rocks, and spoke of the importance of tradition, and how the immanence of the land encouraged it. 'It's worth remembering that those who moved the stones were probably teenagers.' Roger took off his hat and turned it in his hand, as if inspecting it at a milliner's. 'Think of the demographics. Thirty would have been old age, and most never even reached that. Your mother would have been a young teenager when you were born and, if she lived until you were ten years old, you were lucky. Only those children who learned quickly the craft of living would have survived – if you didn't heed the warnings about not wearing wet clothes for instance, you risked your life. With such brief generations and harsh conditions, there's no room for experiments or rebellion. Tradition represented the sort of practical wisdom that kept you alive.'

'Is that why place was so important, do you think – because it was fixed?'

He nodded. 'Look, show me your map. You see where Rough Tor is in relation to Garrow?' I placed a pencil along the line between the top of Garrow and the peak of Rough Tor. It ran exactly parallel with the northings. We were sitting due south of Rough Tor.

Roger had sent me the draft of a paper he was working on with one of the county archaeologists that explored his theory about orientation. Almost all of the stone circles, he said, hold Rough Tor to the north – though the Hurlers refer to Stowe's Hill.

'The reason,' he said now, 'is the Pole Star – the one thing in the sky that does not appear to move. Imagine being here at night and seeing it sit exactly over that hill – the most dominant piece of the world that you know. The whole heavens are shifting – except that one star, above the silhouette. For a people with few certainties,' he continued, 'the power of the Pole Star and Rough Tor must have been phenomenal.'

We returned to the top of the tor and retraced the approach to the 'sacralised' area. I'd walked this way often that summer. To begin with, Rough Tor is in sight, but as you draw closer to the last section, up a slight incline, the view is blocked by the shoulder of the hill and by rocks. On reaching the 'entrance', Rough Tor appears again. It is always a dramatic moment. We paused there, and as we stood, we both noticed something – a group of granite uprights placed in cracks in the bedrock. It looked like the remnants of a wall, running across the threshold. We examined it more closely. It was clear the uprights were at right-angles to the plane of the rock, that they had been inserted.

Roger was animated. 'It would hold back the view of Rough Tor – then, suddenly, you'd see it!'

We examined the base of it for weathering, or lichen growth at the joints. The closer we looked, the more unnatural the insertions appeared, and they weren't recent. We walked down, chattering with excitement.

'All those times up here, and I'd never seen it!' Roger spoke about his work at the Hurlers, and how the three stone circles there might reflect three different eras of use, as the Pole Star shifted its alignment with Stowe's Pound; and about a possible Bronze Age processional route he had stumbled on. I was struck – not for the first time – by the deceptive stillness of the moor, so inert-looking and yet constantly throwing up something new. We passed some low banks. Casually, he said, 'Clearance cairns. Pete Herring thinks they're clearance cairns. He thinks that here might be the very start of farming up on the moor.'

Roger suddenly stood still. 'What we're looking at – all these things, they're just an indication of what was *valued*. Each of these alterations is evidence of a choice, or a series of choices.'

'But why did they bother? What motivated them?'

He responded with a half-smile, and we fell back into stride. 'Maybe it all just came from what you and I are doing – enjoying the landscape and trying to work out what on earth had gone on here.'

It occurred to me then that we've come full circle. Perhaps the reason we're able to 'see' ritual landscape is because we're now less constrained by our own doctrine, by centuries-old habits of thinking in terms of space and not place.

We reached his car. 'I wanted to show you something, but we haven't had time.' He drew from a plastic envelope a folded sheet of A3. It was the photocopy of a map of the whole of central Cornwall, marked with dozens and dozens of dots. 'These are barrows, burial cairns. Do you notice anything about them?'

'No.'

'No, you can't really, not from this. But almost all of them, at least in central Cornwall, are positioned within sight of this – ' he tapped his index finger to the north of the moor – 'Rough Tor.'

I frowned. Aligning sites on the moor was one thing, but over half of Cornwall?

'These clusters here, along the ridges, and on these slopes. And here, down at Taphouse…' Sensing I was slow to be persuaded, he pulled out another sheet. 'Look – ' He had shaded areas where Rough Tor was *not* visible, the Camel valley for instance. Sure enough, there were few dots in these 'view-shadows'. Somehow the negative evidence was more convincing.

'What about these?' I pointed to the ones further west.

'When you get here, these cairns begin to refer to Carn Brea, and then further west –' His hand left the map and swept over the bonnet of his car. 'They point to Carn Galver. It's not all of them, by any means – but enough not to be coincidence. I've visited almost every one.'

I looked more closely and saw the tumulus in our field. 'This one?'

'Yes, I came down to your place a few years back. There was no one there so I poked around.'

'But you can't see Rough Tor – I've checked.'

'Not Rough Tor – but maybe Hensbarrow, before that whole area became covered with china-clay mounds.'

We said goodbye and I watched his car reverse back up on to the open grass verge. Then I set off back into the moor. Hearing him call, I turned. He was leaning out of his window. 'We'll meet up here again. We'll walk, and talk non-stop!'

We did meet once again, and walked and talked and promised each other to meet next time near Wadebridge, to examine some of the more distant alignments from his map. But the months passed and the building work pulled my focus away from the moor and the past. Then I had a phone call from a mutual friend: Roger had become ill quite suddenly. He had died soon afterwards. His great work on Cornwall's prehistory was still incomplete.

# 5 | ROUGH TOR

Old English: *ruh torr*, 'rough outcrop'. The earliest record (Roghetorr, Rowetorr, 1284) coincides both with the approximate date of the loss of the Cornish language in this area and with the abandonment of much of the higher ground; any earlier name for the hill has been lost.

It was early September. I was purging ivy down at the creek. The waxy-leaved billows had smothered and killed so much hawthorn and blackthorn, and were so thick on the oak and holly, that I felt duty-bound to get in there and fight the trees' corner with my pruning saw. I spared the majority of the ivy – the umbels abuzz with hoverflies and bees and wasps were proof of how much their flowers are needed by insects in early autumn. But a good deal of the big ones had to go. I crawled in, sliced through the stalks, dragged them out and piled them on the foreshore.

A by-product of this brutal intervention was that it revealed unnoticed sections of micro-topography – an inlet and a spur, an old pathway. In a couple of places, the ivy had covered small quarry cuts, and I liked to think I recognised some of the bedrock – its grain and colour the same as the stonework of the house.

The days were close and sultry; it was hot work. On the first evening I rested beneath a favourite oak, lay back on the low cliff and let the stillness of the creek settle over me... Suddenly all around, all at once, in hair, clothes – *wasps*! A nest – right underneath me. The drop to the mud was about three metres. I was already down there, ripping off my shirt to let them out, flicking them from my head. Fifteen stings! (Was it revenge?)

I took no more chances on the bank, and the following evening waited for the tide and rowed the boat out into the channel, dropped anchor and slipped into the water. Swimming up the side creek, I felt the last of the flood speed my progress between the trees.

A strange thing about the creeks up here is how difficult they are to describe – not in the sense of conjuring up the mood, but simply being able to refer to them. They are almost entirely devoid of names. Long ago, when the hamlet was still a hamlet and the trading ships still beached on the shingle and the mussel beds were thick and harvested and the ferry went up-river to Ruan, every pool and mudbank would have had a name. Then the sediments settled over those activities and the people left and they took the names with them. Something similar happened on Bodmin Moor: the tin-streamers and stockmen of the Middle Ages used names that are now forgotten, and survive only in the old Stannary Court Rolls and the Launceston Cartulary.

We found ourselves having to make up names, like presumptuous Victorian explorers. The old quay that I'd found on an eighteenth-century map was long gone; when we'd cleared its cutting and began to use it, it became the Old Quay. The headland opposite became Pascoe's Point (after the farmer who owned it), likewise Pascoe's Wood, which covered it. Across the main channel was a steep oak-thick slope that curved round with the river's great northerly bend, and another little headland where the egrets stood in the trees, and that was Penkevil Point (after a village over the back of the hill). Boat expeditions produced rather more visual names – Wreck Creek (an inlet where an abandoned yacht leaned over against the trees), Samphire Creek (thick with marsh samphire), Mangle Bank (where an old mangle had been dumped over the cliff) and Stick Island (a mud island where a stick had stuck).

The names we used were all basic, functional; evocation did not come into it. But traditional place-naming was the same. The most common form in Cornish, as in English, is a compound of two elements, a feature and a description, a generic and a qualifier (our names at the creek were all like this). The main difference between the two languages is the order. English names tend to have the qualifier first ('new-castle', 'ring-wood'), while the Cornish puts it second. Thus in Cornish, frequent place-elements are *pol-* ('pit' or 'pool'), *bos-* ('dwelling', 'cottage'), *tre-* ('farm'), *pen-* ('head', 'end'), *bal-* ('mine'), and typical adjectives are *-dhu* ('black'), *-vean* ('small') or *-noweth* ('new'). Poldhu, Baldhu, Pendhu are all real places, while there are numerous Trenoweths.

If language use is ninety-five per cent literal and five per cent poetic, or some such heavily skewed proportion, then place-naming reflects it. Brevity lends to many place names a certain cryptic pleasure (in Cornwall: Galowras – 'ford of light', or Chytane – 'house of fire'), but most toponyms derive from the simple need to communicate, to identify location. The 'where' is among the first elements of all human exchange. Most stories, most commands, myths and jokes are meaningless without the establishing of place.

One group of Cornish names, though, is more evocative. They are not the oldest by any means, nor do they offer a glimpse of some dazzling Celtic faculty for metaphor. They are instead the bastardised progeny of roving Englishmen. A small headland at Newquay was once *Lost an glaze,* Cornish for the plain 'grey-green promontory'. It is better known now in English by the more suggestive Lusty Glaze. *Scawgack,* meaning simply 'place of elder trees', becomes Skewjack; *Splat an Redan,* 'bracken plot', is Splattenridden. The tendency in Cornish for syllables to be elided adds to the mutations. A cove near Land's End was originally known as *Porth-east* ('Just's Cove'), eroded first to *por-east* and from there to *per-east.* On maps it is now Priest's Cove.

The Manacles are a group of rocks south of Falmouth Bay whose name in English carries a good deal of the menace they represent for ships; they are among the most dangerous and destructive places on the entire British coast. But 'manacles' is a probable corruption of the much less threatening and more prosaic Cornish *maen eglos,* 'rocks of the church' (the steeple of St Keverne is visible from the rocks). It is not hard to imagine the moment – a Cornish pilot guiding an English ship in past the rocks, or an Admiralty chart-maker leaning over the rail to ask the name. In Brian Friel's 1982 play *Translations*, a pair of Royal Engineers have been sent to a small Irish village to map the area, to jot down place names in a similar way – Bun na hAbhann ('mouth of the river') is rendered Burnsfoot. Such bluff anglicising happened wherever the English went.

A few miles from our house is a Texaco garage and Londis store and a couple of houses. They are located at a main junction and the name of the place – Bessy Beneath – is widely thought to tell the story of Bessy, a witch who was hanged and buried at the crossroads. But it is in translation that the story grew, as if only in the no-man's-land between languages can such imaginings take root. The original was probably the Cornish *bos-veneth*, 'small dwelling'.

At high water, the tide slackened and I swam up into the last pool of the creek. The lowest branches of the oaks were underwater; I felt as if I was entering a flooded forest. Beneath the surface, I could see the leaves' fish-like flashing. Sweeping aside a branch, I pulled myself in underneath a group of overhanging trees and stood there, chest-deep.

Out of the distance came the sound of Canada geese. There'd been more in the last week, now that the days were growing shorter. Fifty or sixty – black shapes beyond the leaves, with their heavy flight. And the noise! Hooting and honking at each other as if they couldn't agree where to land: *Here, not here – there, not there, here, not here*... until a calm spread through them and the first few stilled their wings and glided down towards the water... But no. They were rising again, crying, *Not here, not here – there, there, there, not there* – circling around out of sight and their cries fell again until there was just one croaking: *Here, here, here.* With a cymbal-like splash, I heard them land on the water.

I remained a while longer in that leafy capsule, then pushed back downstream with the tide. It was the last swim of the year.

A few weeks later, the building works began. We moved out and, with the season over, were able to rent one of our neighbour's holiday cottages, formerly the farm's pig barns. It was close enough to our house to hitch an extension to the phone line and bring it down via the coppiced limes and in through the bedroom window. By day, vans came and went – builders and plumbers and electricians, builders' merchants, builders' friends, planning and regs officers, heating engineers, bat wardens and insulators. Walls were pushed through, floors pulled up, partitions demolished. Truncated wires and pipes hung from voids like blood vessels mid-transplant. The house became alien to us, an anaesthetised patient in the hands of surgeons, and we could only sit and wait while its insides lay exposed.

One week in November, it rained hard for four or five days, then on Saturday it rained even harder. Water poured off saturated ground, down the track and into the yard below the cottage. Outside the door we watched the growth of a small brown lake across the gravel, and the level nudge towards the threshold. With our neighbours, we grabbed shovels. We dug a channel. It was like the game on the beach with freshwater streams, digging to divert the flow through the sand; only now it was for real. The gushing sound from the bank above was unnerving as we worked – flowing water where no water should be. We dug and dug. The level rose. The fire brigade arrived and the night flashed blue and with the throbbing of their pump and their shovels added, the rush of water was steered away from the house, and on into the creek. The pool around the cottage began to drop. It was still raining heavily.

I went up to check our own house. The rain streaked through

the beam of my head torch and pattered on my waterproofs. I could hear water cascading through the guttering, spitting from a piece of broken downpipe. I walked round the back of the house. No ominous squelch. Inside, in the kitchen, the bare earth of the floors was as crumbly as when we first lifted the flagstones. Once again I thought of those who'd built the house, who'd 'dwelt' here truly and authentically, and I felt a pit-of-the-stomach admiration for their knowledge of just how the land drained, even *in extremis*. They probably didn't even think twice about where to site the house.

In mid-December, the wind went round to the east. Weeks of cold set in. A barn owl took to hunting by day across the paddock, white as a ghost in the dim winter light, quartering the field on silent wings. A midden of building rubble now half-blocked the front of the house – twisted metal and concrete and broken timber, sandwich packets and Coke cans. Brushed with frost it now looked a little better. There wasn't much snow, but when it did come, light and dry one dusk after everyone had left, it blew in through the kitchen door where I was standing waist deep in a pit. I watched the flakes drift down like feathers, to rest on the bare earth, on the muddy toe of my boots – unmelting. In that moment I found it hard to imagine the house ever being habitable again.

But at the beginning of January one of the plumbers stood up beside the boiler, wiped his fingers on a rag, and said: 'Well, that's us about done.' The electricians were nearly finished, the wood-burners were in, the structural work was done. The plasterers arrived and strapped on stilts to skim the ceiling, and we were talking about dates just a month or so away for moving back in.

With the flood of daily decisions reduced, my horizons expanded. Somewhere, on the dressing-table in the rented barn, beneath the plans and delivery notes and invoices, the scribbled dimensions, the slate samples, a child's sock, the proofs for the new book, were last summer's notes and folders. I had been gathering maps and references and books and off-prints of academic articles and now had a route of sorts – down from Bodmin Moor to Tintagel and along the coast, over to Goss Moor and the clay country, then down the banks of the Fal, and on into the far west. In parallel, there was the age-by-age story of topography – from the sacred landscapes of the Neolithic to the Middle Ages, the Enlightenment and into our own post-industrial age. Also, scattered through my notes was a half-assembled cast of topophiles from various centuries. As the house and other commitments allowed, I would follow the route.

The first chance was a week or so in early February. I headed back up to Bodmin Moor. My cousin Richard was staying with us and he came too, with his mittens and his sketch pad and a sheepskin hat he'd bought in Damascus, having walked there from Istanbul just before the war. Richard had been walking for years, half his life, and painting as he walked, and when he wasn't walking and painting, he was living and painting in a barn in central France with hardly any walls. I didn't think he'd mind the moor in winter.

But it was cold – the coldest day of the year. Five below was forecast and it was already down towards freezing when we tramped out across the moor in the mid-afternoon. We were staying in another of Julie's abandoned buildings. Priest's Hill is single-storied and single-roomed, like an old longhouse. It is exposed and desolate and older maps show it as Mount Pleasant, a sarcastic name given by the miners who were billeted there. Expanding foam had been sprayed into gaps in the slates, but the wind still poured through. We sawed up logs, laid out our stores and, with just enough daylight, hurried off to climb Garrow.

The grass was already crunchy with frost. We passed the hut circles and old field boundaries and came out on top. Approaching the 'threshold' at the northern end, seeing Rough Tor disappear and come into view was more striking now, in winter. The rocky peak stood proud in the fading light, rising from the black plinth of its own shadow. The late sun glowed to our left, absurdly big in the reddish haze. As we watched, tiny black spots appeared against it. They looked like flies on a plate-glass window. But they were starlings, massing, heading north. It took a long time for them all to cross the sun. They pushed on into the half-darkness, around the back of Rough Tor, not flying straight but flowing like water, no longer individuals but moving as one, swaying, dipping and rising and tumbling beside the stillness of the tor.

The next day, I woke before dawn. The fire had long since gone out. Richard was just a sheepskin hat at the top of his sleeping bag. Pulling on a coat, I stepped outside for a piss, and at once found myself stumbling around like a drunkard. The cattle-hoofed mud, soft last night, had frozen solid. Staggering back to the lean-to, I gathered wood and stood to face Rough Tor. Silhouetted against the night sky, it looked like a moored boat and I found myself checking its position as if it was my own, to see if it had dragged its anchor. It was lying well, moored beneath the Pole Star.

I hurried inside, made a little cone in the grate from the kindling I'd brought (old laths gathered from the rubble at home), added the logs and burrowed back into my sleeping bag. The flames rose, spitting and cracking as they caught the century-and-a-half-old wood. Soon I could feel warmth on my face. All those years for a moment's relief! Rough Tor was just visible through the small window, perfectly framed. The stars above it were fading now, and the sky was growing pale. I read for a while and when I looked up, the first rays of the sun were on the highest of its rocks, gilding them.

I'd been driving a good deal around Cornwall, sourcing and collecting materials for the house. Cresting a ridge or coming round the back of a hill, I played the game of trying to predict whether and where Rough Tor would be in view. When it was – however partial, however distant – its presence appeared to fill the entire view; and when it wasn't, there was a sense of unease, of expectation.

I remembered the same feeling in Yerevan, where the snow-capped presence of Mount Ararat dominates the city, so much so that when it was hidden behind a block, I found myself walking faster to see it again. I could feel it at my shoulder when it was behind me. In cafes, if it was visible through the window, I noticed that people would arrange themselves in unconscious homage to it. 'I have cultivated in myself a sixth sense,' wrote Osip Mandelstam in Yerevan in the 1930s, 'an "Ararat" sense: the sense of attraction to a mountain.'

Sacred mountains crop up in most traditional cosmologies. Mesopotamia had the 'mountain of the lands'. In Asia, the mythical Mount Meru, subject of so many cosmographic images, was believed to revolve around the axis of the Pole Star. In earthly terms Meru is usually equated with Mount Kailas, each of whose four sides – north, west, east and south – produces a great river. Herodotus wrote of the Zoroastrian god, Ahura-Mazda: 'In their system, [he] is the whole circle of the heavens, and they sacrifice to him from the tops of mountains.' Like Muhammad, Zoroaster received his divine revelation on a mountain – Mount Ushi-Darena – and in the Zoroastrian sacred text, the *Yasna,* a phrase about the sanctity of mountains is repeated: 'all mountains that shine with holiness, with abundant brilliance, Mazda-made, the holy lords of the ritual order.' Olympus, Tabor, Sinai, Ararat, Fuji, T'ai Shan, Machu Picchu . . . it's hard to think of a great mountain that is not linked with the gods, or even a distinctive hill that has not at some stage generated a local belief.

Some years ago, as part of a longer journey across Mongolia, I visited the sacred mountain of Otgen Tenger. In Ulan Bator a friend had introduced me to Narmandakh, a student of tourism who, like most young Mongolians, had spent his summers herding and riding. We travelled down to the town of Uliastai and bought three horses, from three different families in three different *gers*, or tents: one horse each, and a packhorse. Days before reaching Otgen Tenger, we first saw its snowy dome suddenly floating above a ridge, high in the cloudless blue. It would come and go as we rode – sometimes hovering like the moon, sometimes absent, leaving us searching the horizon. We stayed below the mountain with a herding family. For two days it rained, but on the third day the sun rose in a clear sky and we rode up to the mountain with our hosts. The closest to the peak anyone was permitted was a lake just below. I remember the mountain's upside-down reflection in the water and the water itself like a sheet of silver; and on the shoreline, and rising from the water itself, dozens of cairns. The oldest of the herders prayed. A mile or so back, before wading the river, he had dismounted and turned away, calling over his shoulder: 'Do your piss here! After this it's sacred land.'

What left its mark on me from those weeks was less Otgen Tenger itself than its part in a cosmology in which everything, every aspect of life, was integrated. The land we rode through by day was mirrored in the *gers* we slept in at night. The blue dome of the sky became the white dome of the *gers'* ceilings. The night sky, in Mongolian cosmology, is the *ger* of heaven, and the Pole Star is the 'nail star', pegging the heavens to the ground. The day's sky is in the blue of the sacred scarves – flicking from *ovoos* in windy places, and hung from the necks of cattle for protection – and the holy mountain Otgen Tenger means 'younger sister of the sky'.

'*Tenger*' isn't just the word for 'sky'. It is also the word for 'weather'; in fact there are ninety-nine *tengers*, the ninety-nine names for the gods. Nominally Buddhist – Tibetan Buddhists of the Gelugpa school – Mongolian herders have never lost touch with their own deities, their *tengerism*. They haven't migrated or altered their lifestyle very much in thousands of years. Generations have driven their herds through the same land, seen the same knolls and rocky outcrops, the same pebble-bottomed rivers, the same lone trees, beneath the same peaks. Places are an assembly of gods and ancestral spirits, an animistic map of embedded memory. Listen to their talk and the name of each location brings out a story or recollection. Ask your way and you receive a litany of landmarks – pass the hill like a woman's breast, the rock like a rabbit, the ridge like the neck of a horse.

Never have I felt so aware of orientation as during those weeks. Rather than floundering in a surfeit of space, with few physical features and the knowledge that thousands of miles of land stretch out on all sides, I found the cardinal points forming an invisible grid around me. The *ger* itself is a microcosm. You always enter from the south. You duck through a low wooden door. You place your saddle to the left and slip off your boots. You carry on clockwise into the tent. In the middle section, to the left of its central axis, you kneel on one knee and accept – always with the right hand – the bowl of tea held out towards you. To your left – always the north – is the family shrine, a brightly coloured chest containing the treasured possessions of the family. On top of the chest are displayed black-and-white pictures of dead grandparents, favoured lamas, the Potala in Lhasa, or one of the Bodhisattvas. (As we drew closer to Otgen Tenger, so pictures of its peak would appear more often in this gallery.) You give a respectful half-nod to these images before sipping at your tea. Later you will spread out your bedding and, when you wake, you will look up at the smoke-hole at the tent's apex, the *tono,* to see whether the stars are still shining or whether the first pale of the day has reached the sky.

Now, years later, reading around the idea of ritual landscape, it struck me how frequently places and mythologies are interchangeable. In 1926 Bronislaw Malinowski set out his functionalist definition of myth:

> I maintain that there exists a special class of stories, regarded as sacred, embodied in rituals, morals and social organisation, and which form an integral and active part of... culture. These stories... are to the [people] a statement of a primeval, greater, and more relevant reality by which the present life, fates, and activities of mankind are determined.

Replace 'stories' with 'places', and the definition works just as well. In the animist past or in pastoralist societies like the Mongolian herders, the land is a story-book, places are narrative, and high ground is the home of heroes and gods.

For the first hours of that morning Richard and I sloped about the hut in a state of inner-chill and sleeplessness. We cooked, we cleared up, then ambled down to the stream. The sitka plantation shielded the wind and, with the sun climbing higher, a warmth entered that sylvan arena and seeped through to our bones. Richard sat sketching near the bridge. I wandered upstream and perched on a boulder to let the sun do its work. I could feel it on my shoulders and see its reflection flickering on the underside of the bank. A crowd of bubbles bobbed in the backwater until the flow caught them and sped them over the rocks. The plantation enclosed the sound of the stream and the passerine song high in the trees. A buzzard launched itself from inside the canopy, took two beats, and found some hidden camber on which to glide. I was suddenly filled with optimism about the coming spring, the westward walk ahead and more such loafing beside streams.

Back at Priest's Hill we gathered our stuff and locked up. To the north the moor glowed gold as we started across it, towards the sun-lit shape of Rough Tor. The sound of an engine made us turn and there was Julie on her quad with two fire-beating poles strapped to the back. She was going up to Garrow to burn the gorse. 'One match – ' she smiled – 'and whoosh!'

We went our separate ways. A little way on was Fernacre Farm, the only moor-farm fully surrounded by the commons. Its squares of pasture looked unnaturally green against the pale of the downland, the sheep on it unnaturally white. The farmhouse was half-hidden behind the hill but I could see the turning blades of a turbine. Julie's brother had farmed Fernacre until a couple of years ago; everyone I spoke to said what a wonderful man he was. But he was killed in a shooting accident. Looking now at his land, I suddenly recalled something an archaeologist had said to me about the Neolithic: 'Imagine how depleted your people would be after a harsh winter. Imagine the spirit of loss blowing across the moor.'

A mile away behind us now, I could see the dot of Julie's quad-bike on the slopes of Garrow, and beside it, in the windless air, a steeple of granite-grey smoke.

The approach to Rough Tor started well enough. We crunched swiftly over King Arthur's Downs. The winter sun filled the moor and Rough Tor filled the sky ahead. Then we cut across a low valley, the ground softened and we were up to our calves in bog. We looked at each other, thirty metres of wet ground between us. 'Ah!' called Richard. 'Right!' I called back. The logic is always to go on – hopping boldly from tussock to tussock, even as the tussocks grow further apart, as they quiver at your footfall. It is a logic that should be firmly resisted. *Retrace your steps! Take the long way!*

'Brown Willy and Rough Tor,' wrote the Victorian novelist and moor-wanderer Sabine Baring-Gould, 'are fine hills rising out of really ghastly bogs. Places to which you would hardly desire to consign your best enemies, always excepting the promoters of certain companies. I really should enjoy seeing them flounder there.'

We back-tracked, in the end. Rough Tor's lower slopes swept upwards from the mire. Short winter grass gave way to clitter fields and jumbles of lichened stone. The ground steepened. The rocks grew thicker across the slopes until we were using them as steps, listening to their porcelain *chink* as they wobbled beneath our weight. If you look closely, it is possible amidst the chaos to spot orthostats – rocks which have been propped up or deliberately shifted. And it is possible too to make out the walls of a 'tor enclosure', as at Stowe's Pound. But the one at Rough Tor is much more substantial and elaborate.

H. O'Neill Hencken, who wrote the standard text for Cornwall's archaeology in the 1930s, describes Rough Tor as 'an ancient stone fortress'. Even as late as the 1990s, Johnson and Rose in *Bodmin Moor: An Archaeological Survey* caption their map 'Rough Tor Hill-fort' and speak of the walls as 'ramparts'. But the twentieth century was more martial than our own age, and ritual is now the keyword. Amidst the grey expanse of boulders that ring the peak, the 'walls' are little more than a few raised stones, like hands waving from a sea of stone. In a couple of places are gaps which can be read as gateways and it wasn't hard to see how, with the land dropping away on all sides, it might look like a castle.

Stepping up to the rocky summit, I carried with me a more recent interpretation – the alignments, the dozens of sight-lines converging on the hill like flight paths on an airport. They made a shrine of this rocky hill, a temple, and added reverence to my steps up to it.

A few months before he died, Roger Farnworth had sent me a series of photographs he'd taken. They showed the setting sun in a sequence as he'd followed one of the walkways. He'd written the times beside each photo and noted that at normal walking pace, the dip and rise of the hill meant that the sun appeared to rest on the peak. He had wanted to repeat the photos at the winter solstice: 'We could take photos more methodically at five-minute intervals. A cloudless sky is essential.' When we'd gone up together on 21 December, there was thick cloud. He never had the chance to repeat it.

Of the various features of Rough Tor that Roger was able to show me that day, there was one whose simplicity made it more affecting than the others. Tucked in beneath one of the granite stacks was an overhang and beneath it a cave-like chamber. In the chamber, if you leaned down and looked inside, there appeared in the shadows piles and piles of stones. They were not like the other stones of the summit. Each was small enough to be carried. No frost action or erosion could have left so many. It is tempting to think that at some point – perhaps during the Neolithic or early Bronze Age when the peak was so revered – thousands of supplicants might have carried them up and placed them in those hollows. I thought of the rock-bank at Stowe's Pound and the Mongolian stupas at Otgen Tenger, and how high places still provoke such gestures.

Half way down the western slope of Rough Tor is a grassy bank. It stands a metre or more high and is wide enough for two people to walk abreast. It runs for about four hundred metres. Neither the beginning of the bank nor the end of it marks anything in particular, nor is the ground to one side of it any different from the other side. From the well-worn footpath up Rough Tor, the bank makes little sense. Its only recent use – perhaps the only one for several thousand years – was during the Second World War, when the army cut through the bank for tank-firing practice. But the bank cairn is now considered one of the more impressive of Britain's Neolithic monuments, and one of those that best illustrates the ritual importance of natural features like Rough Tor.

The point of the bank, it is now believed, was simply to walk up it. After the first few metres, it is clear why. Directly ahead, on the skyline, stands Showery Tor, northernmost of Rough Tor's ridge of rock formations and itself neatly ringed by a perimeter of deliberately placed stones. Then the bank curves away – but only to set up a new alignment with the outcrop at Little Rough Tor. Here's this place, the route implies, and now this. Gesturing at two of Rough Tor's gallery of weather-sculpted features embodies much of what we know – or think we know – of this most enigmatic era of our past. Walking it now, as Richard and I did that cold afternoon, is all the more powerful for its plainness.

One of the more surprising aspects of the ritual monuments of the late Neolithic and early Bronze Age is that, after several thousand years, they suddenly ceased to be built. From about 1500 BC, the archaeological record reveals no new ones. Instead there are a profusion of hut circles, field systems and functional artefacts. All that effort constructing banks and hauling rocks for no practical reason, was suddenly abandoned in favour of home improvement. Not for another two and a half thousand years does sacred building appear in Britain on such a scale – when Gothic cathedrals and hundreds of parish churches began to rise from the land.

I waved Richard off on a bus in Camelford and the following afternoon walked down to the coast, through a series of field-edge paths and high-sided lanes. The winter's browns were broken by yellow bursts of gorse and daffodil. The last settlement before the sea was Trevalga. As I drew close, my internal banter was making connections – *Trevalga, Trafalgar, ancient seaways and the Gates of Hercules, Cornish tin and Iberian tin*... But no – 'Trafalgar' is from the Arabic *tarf al-gharbh* ('cape of the west'), while the best bet for 'Trevalga', via second-syllable mutation, is Middle Cornish for the farm of someone called Melga.

The village itself was an oddity, with sagging slate roofs and slate-clad farms and cottages clustered around the church like piglets around a sow. It looked ageless. As I dropped through the deserted lanes and out again down a high-banked track, I wondered if I might have dreamt it. Then – a sudden V of land, and the sea! A swell was running. I could see the low sun shining through the crests with a green-blue translucence that made my spirits soar.

It was well after dark by the time I reached Tintagel. A cold wind drove down Fore Street, stinging my face and rattling the scalloped awning of a gift shop (*Closed for Season*). I was the only guest at the B&B. Dumping my bag there, I walked down the valley towards the beach. I could hear the seas sweeping round into the haven, hissing as they sent their surf far up the shingle to drop back again, knocking the pebbles together like billiard balls.

In the darkness to my left rose a shadow, a rough-edged silhouette cut out of the moonlit sky: the 'island', and a bridge arced from it back to the mainland. From that slender link, that corridor of cliff connecting one rock to another, has grown one of Britain's most enduring stories.

# 6 | TINTAGEL

*Dyn-, tin-, dinas*: as in a number of Cornish place names, usually translated as 'fort' (Irish *dun-*); *-tagell*: 'throat, constriction', from the root verb also in Welsh and Breton *tage*, 'to choke', and referring to the isthmus that connects Tintagel Island to the mainland.

In the Public Record Office in Kew, beneath fierce overhead lights, the leather binding showed its age. The nap was raised where dozens of fingers had opened the book to consult the land charters inside. Among them is one dated May 1233, which goes a little way towards explaining the particular mystique of Tintagel: 'Know ye all men present and future that I Gervase of Tyntagel have given, conceded and quit-claimed from me and my heirs in perpetuity to the Lord Richard the whole of my island of Tintagel.'

'The Lord Richard' was Richard of Cornwall, younger brother of King Henry III, and his revenues from Cornish tin were helping to establish him as one of the richest men not just in England but in Europe. In exchange for the single manor of Bossiney (containing Tintagel), he offered Gervase a total of three others in parts of Cornwall much less harsh and more fertile. An echo of coercion comes from the soil-coloured ink of Gervase's statement, as well as from the terms of the deal, a hint from eight centuries past of a little princely arm-twisting. It is clear Richard was very keen to add Tintagel to his growing list of properties.

He built a castle. On the mainland he raised an impressive set of outer walls. The upper ward rose from a high jut of rock, fringed to the west by cliffs which dropped sheer, hundreds of feet, to the open sea; the lower ward was protected by a great ditch and more cliffs, and led to the Island. The neck of land was higher then but still required a narrow bridge. Walking across it, the visitor could hear the surf below and see the plunge on each side and the gatehouse ahead.

That evening, I looked at the bridge high in the moonlight, with the winter wind at my ears and the Island black against the stars.

How easy it is to let such places pull you into a fug of medieval flummery – clanking knights and unworldly quests, wimpled maids and minstrels, caparisoned horses and sieges in rook-rounded towers. Tintagel Castle is one of many similar buildings – from the Crusader castles of the Levant to the Cathar castles of France – whose lofty position allows us to imagine the horror of medieval warfare, the terrifying privations of a siege.

But Tintagel Castle was a sham. It was a show castle, a Potemkin facade. It was never besieged, never attacked. It protected neither seaway nor landway, provided no sanctuary for soldiery. Richard never depended on it as a personal refuge (he had eight other castles scattered across southern England). He never lived there. In fact, no record survives of him even visiting the place. His reputation, wrote one biographer, was not as a 'great warrior, but a negotiator and schemer of genius'. He is remembered chiefly for his duplicitous allegiances, as a hoarder of treasure and, according to a contemporary chronicle, as 'a great lecher towards all women of whatever profession or condition'.

So why did Richard put so much effort into Tintagel Castle?

Soon after bullying the site from Gervase, Richard went on a Crusade. True to form, he managed to avoid most of the messy business of open conflict. He spent weeks in camp at Jaffa and Acre and Ascalon. His main success came through clever mediation with Egypt's Ayyubid sultan for the release of prisoners. But the Crusades offer a clue to Tintagel's appeal. For several hundred years, the knights of north-western Europe threw their wealth and their lives into recapturing and holding a faraway city of little earthly relevance. It was the noble endeavour of an idealistic age, and one motivated by, among other things, the mythical power of place. Like the bearers of the dead at Aveline's Hole, or those who built monuments around Rough Tor, Richard wanted Tintagel not for its own sake but for the idea of it. He was acquiring its stories and its past.

As Jerusalem's attraction was based on Biblical allusions, so the appeal of Tintagel was textual. In 1136, almost exactly a century before Richard built his Cornish castle, Geoffrey of Monmouth produced the *Historia Regum Britanniae,* the only source for the link between Tintagel and King Arthur. The story as he told it is still well known. At a gathering of nobles in London, Uther Pendragon, king of the Britons, falls for Igerna, the wife of the Duke of Cornwall. As a result the duke storms back to Cornwall with his coveted wife. Uther Pendragon follows with his army. The duke leaves Igerna at Tintagel and marches off to another of his strongholds. The perilous approach to Tintagel means that no military force could possibly gain access. Only by means of Merlin's sorcery does Uther Pendragon succeed: he is transformed into the guise of the Duke of Cornwall and in this way makes the crossing to the Island, and reaches Igerna. Their union produces the child who would grow up to be King Arthur.

In Cornwall, Arthurian passions certainly predate Geoffrey's account. In Bodmin in 1113 (a couple of decades before Geoffrey's *Historia*), a group of visiting French canons scoffed at a local claim that Arthur was still alive – and nearly provoked a riot. But there was no apparent connection between Arthur and Tintagel until Geoffrey's *Historia* claimed it. The growing impact of the book perhaps explains why the family who owned Tintagel changed their name. Until the beginning of the thirteenth century, they were 'de Hornicote', after their principal property. Tintagel was only a small part of the lesser manor of Bossiney – yet by the early thirteenth century, they were styling themselves 'de Tyntagel'.

For the Cornish people, the Island and its isthmus did have an earlier association. Some eight hundred years before, the site appears to have been the seasonal seat of Cornwall's kings. From that time comes an extraordinary set of finds. As part of excavations at Tintagel in the 1990s, archaeologists Charles Thomas and Carl Thorpe examined pottery found on the Island. I visited Charles's home near Truro at the time, and I remember seeing several large Formica tables in the garage spread with dozens of mud-coloured earthenware shards. It looked like the aftermath of a domestic accident, just before the Araldite comes out. Carl was sifting through them. Much of the pottery, it turned out, came from Carthage and the Byzantine provinces of Western Anatolia. 'More pieces from the Eastern Med,' Carl later told me, 'have been found at Tintagel than in the rest of Western Europe put together – and only about two to five per cent of the site has been dug.'

The suggestion is that after the Romans left Britain, this small rocky projection on the Cornish coast retained a link with the Byzantine world more important than any other known site in the country. Tintagel's value as a harbour then was greater, with the sea level lower and ships smaller. But I can't help feeling that the Island's prominence might even then have derived from its past, from traditions and stories that arose from its spectacular position.

The number of places in Britain associated with Arthur is vast. More than a hundred toponyms indicate where he lived, dined, sat, slept, hunted, fought, was buried, left his thumb-mark in stone or allowed his dogs to drink. They occur particularly in the west, in Cornwall, Wales and southern Scotland, and on the rocky outcrops, hill-tops and moors. Oliver Padel concludes that the Arthur of these places was not so much the *dux bellorum* of the written record – the victor at Mount Badon, the ancient British monarch embellished by Geoffrey of Monmouth – as a generic mythical figure who goes back much further than the post-Roman centuries. 'The cumulative weight of early evidence,' writes Padel, 'indicates that [Arthur] was primarily a pan-Brittonic figure of local wonderment.'

It is usually assumed that Arthur's stature encouraged the naming of places after him – first the story, then the place. But Padel's idea reverses the pattern: the figure of Arthur as an aggregate of hundreds of site-specific tales. Somewhere like myth-rich Tintagel would thus have begun with the suggestive topography, the almost-island in which settlers saw the imagined home of a hero or the embodiment of a cherished cosmology. Idea and place became entwined. Later on, a myth-maker like Geoffrey of Monmouth might gather the stories together, elevating local heroes to something more universal, taking them away from their original setting to broaden their appeal. But because we all love places and pilgrimages, the literature sends us out again, to seek out the old hero-sites high in the hills or on the wild coast. And so the stories bounce down through the centuries, their edges rounding and their shape altering with the collective spirit of each age.

When Richard of Cornwall built Tintagel castle in the thirteenth century, there may still have been folk memories of Byzantine ships or Cornish kings or prehistoric ritual. But they've all gone now. A stroll through Tintagel nearly eight hundred years later confirms there is really only one show in town. Here is Arthur's Bookshop, King Arthur's Arms Inn, Camelot Amusements and Camelot Castle Gallery. Overlooking the ruins of Richard's medieval castle is the vast eighty-room hotel, built in 1899 as King Arthur's Castle, with battlements and faux lancet-windows, for the growing number of pilgrims encouraged particularly by Tennyson's *Idylls of the King*. Tennyson reignited the cult around Tintagel, encouraging dozens of others – Thomas Hardy, Walter Swinburne, Edward Elgar and Rudolf Steiner – to visit and record their own impressions. In the 1920s, Frederick Glasscock retired to Tintagel with a fortune made in the custard business. He dedicated the rest of his life to establishing a new chivalric order. He built the spectacular Arthur's Great Halls in Fore Street and performed mock ceremonies to make knights of the more worthy of his followers. With some twenty thousand members, the order caught on in America. In July 1934 Glasscock boarded a ship for New York but he died en route and was buried at sea. His Great Halls remain open to the public, and in their church-like interior I watched the winter sun glow in the stained glass and on the stone throne, where Glasscock himself presided, Arthur-like.

The Arthurian tales have more recently brought other camp-followers to Tintagel, the sutlers of crystals and wind-chimes, the tarot-readers and soul-curers. Why they should go together is hard to say. It's as if a portal has been opened in this coastal village, allowing access to all the fantastic possibilities not only of the past but of ourselves. I paused for a moment to read the sign in an empty shop window: *Healing Realms Ltd – COMING SOON! Usui Reiki Healer, Master / Teacher, Thermal Auricular Therapist and Psychic Development Teacher*. A woman with crimson-streaked hair stood beside me and read it too. 'Oh, *good*!' she said, and with a happy smile strode off into the morning.

It is popular among those immune to such therapies to point out that the great edifice of Tintagel – all the visitors and all the shops, the healers and seekers, the generations of myth-weavers, the entire industry of Arthur's birthplace – is built on sand. Geoffrey of Monmouth is hardly a reliable source, and he mentions Tintagel only four times. At no point does he say Arthur was born here, only that he was conceived here. But that ignores the place itself. We can scoff at the centuries of credulity, the castle raised on dubious legend, at all the pedlars of mist-wrapped history. Yet strip back the story-telling and you are left with rock, with the topography. 'It is situate on the sea,' explains Geoffrey of Monmouth, 'and is on every side encompassed thereby, nor none other entrance is there save such as a narrow rock doth furnish, which three armed knights could hold against thee, albeit thou wert standing there with the whole realm of Britain beside thee.'

I crossed the narrow bridge that morning, from the mainland to the Island. I wandered the outer cliffs in that detached state of mind that islands always induce, even one little bigger than a couple of football fields. 'The sea elevates these few acres,' wrote Adam Nicolson of the Hebridean Shiants, 'into something they would never be if hidden in the mass of the mainland.' At the northern cliffs, there was only the sea below and in front, and miles and miles of water stretching out to a dark line of cloud. I looked down over the haven, at the swells swinging in around the point, and thought idly of ships from Byzantium lying there at anchor, and venturers rowing ashore with amphorae of olive oil and news of wars with the Persians, or the latest Christological debates from the eastern Sees.

As I left, climbing up from the castle's upper ward, on past the church of Saint Matheriana and out along the coast, I turned to look back. The Island squatted solidly above the water, but the castle walls were crumbling. They'd been crumbling more or less since the day they were built. Likewise the cliff linking the Island has been steadily falling into the sea for centuries. In recent years, English Heritage has been pouring concrete over it to try and stem its collapse. Visitors now cross it via a walkway protected from the drop by a robust hand-rail. Their numbers show no sign of slowing, nor does the commerce they generate in the village.

Looked at from a distance, the frailty of the link to the mainland is acute. How long will it last? How long before Tintagel Island becomes a real island, one of those rocks around the Cornish coast that know only the feet of fulmars and gulls and black shags, and whose top in summer is spongy with thrift and whose lower rocks even on a calm day are cordoned with white spume? And would its severing mean the end of the Arthur industry at Tintagel, or simply add another layer to the appeal of a place that allows us to float free, adrift on our own sea of stories?

For the rest of the day, I followed the lonely stretch of coast between Tintagel and Port Isaac. The clifftop path wove through a mass of old slate quarries, worked-out dells, blasted rock faces and single standing columns, which looked like the chimneys of bombed houses. I stayed the night in Port Isaac, a pinch-roomed B&B of marshmallow pinkness, and carried on reading Geoffrey's *History of the Kings of Britain* – the Everyman edition, coat-pocket sized and perfect for pulling out on the path.

Geoffrey of Monmouth was the inventor of British history, in the sense that no one else has produced an account of the nation's origins that has been so influential for quite so long, nor constructed such a compelling narrative for the British to explain their distinctiveness. Also, he made a lot of it up. He claimed that his account was merely the translation of a document discovered in Brittany and given to him by Walter, Archdeacon of Oxford. In all probability, that's nonsense. He inflated stories gathered from many sources, stories that had already been distorted from the shape of true witness by years of retelling. He borrowed from other authors and let his own language and imaginings run wild.

Yet his stories were believed. They were read, discussed and quoted – so much so that they seeped deep into the fabric of national identity. In the 1950s, the antiquarian Sir Thomas Kendrick assessed the early impact of the *History of the Kings of Britain*:

> Within fifteen years of its publication not to have read it was a matter of reproach; it became a respected text book of the Middle Ages; it was incorporated in chronicle after chronicle; it was turned

> into poetry; it swept away opposition with the ruthless force of a great epic; its precedents were quoted in Parliament; two Kings of England used it in support of their dominion over Scotland.

There were always sceptics, but to others the text was as unassailable as Genesis. Under Elizabeth I, Geoffrey's work became something of a cult and a number of plays were staged based on it. Even towards the end of the seventeenth century, John Milton could refer to Brutus, grandson of the Trojan Aeneas and Geoffrey's supposed founder of Britain: 'There to thy sons another *Troy* shall rise / And kings be born of thee, whose dredded might / Shall aw the World, and Conquer Nations bold.'

Geoffrey's *History of the Kings of Britain* did in letters what the monuments of the Neolithic did in stone: each drew on an oral tradition, which then disappeared. The stories of King Arthur – almost a fifth of Geoffrey's text – did more to establish Arthur's place in British mythology than any other single work. And there is no disputing that Geoffrey told a good yarn. He told of the glories of the past, of a great dynasty of ninety-nine kings that went back to the twelfth century before Christ (just a few centuries, as it happens, after the sudden cessation of ritual monument-building); he told of a united Britain and an ancient Arthurian empire that pushed up into Scandinavia. He produced the nation's historical charter, its myth of origin, its narrative foundation. It included the millenarian prediction that when the Britons obtained the lost relics of Arthur, the ancient kingdom would be restored. Many streams of motivation fed into Geoffrey's work (personal ambition and hatred of the Welsh were among the stronger), but above all he was producing a pedigree that would help offset the repeated ignominy of the island's invasion by Normans and Saxons, Vikings and Romans.

For Geoffrey was not just the inventor of British history but of 'the British History', the long-held belief that Britons can trace a line of heredity back through a series of wise and foolish kings, through wars with Romans and Gauls and Danes, to a group of ship-borne drifters from the Mediterranean. They were led by Brutus, who sailed to Italy after the end of the Trojan war. By accident, Brutus killed his father while out hunting and was driven from his land, so beginning a series of wanderings that ended on the shores of the island of Albion, at Totnes. Brutus and his followers found the island uninhabited, save for a tribe of giants, which they soon chased off to the fringes. The language that Brutus and his settlers spoke was 'Trojan or crooked Greek'. They established a city on the banks of the Thames and called it New Troy. They named the island Britain after Brutus.

This is the story Geoffrey told. For hundreds of years, it was what a great number of Britons believed about the nation's beginnings. What struck me reading Geoffrey's work in full was not so much the myth-making, the battles and prophecies, as the emphasis he puts on place, on Britain's natural virtues. He relishes the description of magical lakes, fish-rich rivers, fruitful fields. He even chooses to open the book with them. After a dedicatory chapter, but before the appearance of Brutus, there is a lyrical overture to the island, suggesting that the nation's destined greatness grew out of the land's beauty and fecundity. Having built up an impression of the wealth of the country's minerals and the fertility of its soils, Geoffrey delivers a remarkable pastoral passage:

> Forests also hath she filled with every manner of wild animal, in the glades whereof grow grass that the cattle may find meet change of pasture, and flowers of many colours that do proffer their honey unto the bees that flit ever busily about them. Meadows hath she, set in pleasant places, green at the foot of misty mountains, wherein be

> sparkling wellsprings clear and bright, flowing forth with a gentle whispering ripple in shining streams that sing sweet lullaby unto them that lie upon their banks.

For much of Geoffrey's work there is no earlier written source, so whether he invented the tales or merely embellished them from oral traditions, we will never know. In his rhapsodies about the land, however, there are obvious borrowings. A passage in Gildas's sixth-century *The Ruin of Britain* reveals the source of Geoffrey's riparian image: 'brilliant rivers that glide with gentle murmur, guaranteeing sweet sleep for those who lie upon their banks.' And, two hundred years later, in the eighth century, Bede starts his *Ecclesiastical History* with a passage of striking similarity:

> Britain is rich in grain and trees, and is well adapted for feeding cattle and beasts of burden. It also produces vines in some places, and has plenty of land and water fowl of divers sorts; it is remarkable also for rivers abounding in fish, and plentiful springs. It has the greatest plenty of salmon and eels; seals are also frequently taken, and dolphins, as also whales; besides many sorts of shell-fish...

Bede, Gildas and Geoffrey were all drawing on a long tradition, of the *locus amoenus* – the 'delightful place' – a literary trope which goes back to classical literature, and beyond. A good deal of scholarly discussion surrounds the *locus amoenus* and much of it cites the prescriptive definition by Ernst Robert Curtius, who states drily that its components are usually 'a tree (or several trees), a meadow, and a spring or brook. Bird-song may be added. The most elaborate examples add a breeze.'

Freeing it slightly from such straits, the 'delightful place' can be seen to touch something universal. It is there in the idea of Eden (long believed to come from the Hebrew *and*, 'delight') and to the sacred groves and *kunds* ('pools') of Krishna's childhood, the banyan tree of the Buddha and the garden of the Hesperides. The word 'paradise' itself, used in both Christian and Muslim traditions to describe heaven, originates in an old Persian word for 'enclosed garden'.

In a recent study of the *locus amoenus*, Catherine Clarke identifies it as 'one of the most significant rhetorical figures in medieval European literature. The delightful pastoral place or garden re-surfaces in hagiography, lyric, romance, dream vision and drama.' The *locus amoenus* also played its part in the development of English identity. The specifics of Bede and Gildas became embedded in national myth, and Britain's topography could be invoked to assert a collective sense of belonging, as in John of Gaunt's much quoted speech in Shakespeare's *Richard II*: 'This happy breed of men, this little world, / This precious stone set in the silver sea, / ... This blesséd plot, this earth, this realm, this England.'

Clarke notes a recurring motif in idealised places: a sense of enclosure or encirclement – either of the island itself surrounded by sea, or more locally. 'In particular, this English tradition prioritises motifs of pastoral enclosure and containment. Such imagery links descriptions of the island of Britain itself, celebration of local island or hill sites, and representations of English saintly retreats and the monastic cloister.'

The *locus amoenus* generates its power not just as a received idea but because there are few lives that have not been shaped to some extent by the joy to be had from idling in an enclosed or pleasant place. Nature might be hostile, heartless and uncomfortable; it might fill the working day with efforts to cut back growth, keep away predators and pests, the cold or disease. But sitting under a tree by a stream, surrounded by hills, offers not just relief but the chance for those fleeting moments of transcendence that sustain all creeds.

'Dismal is this life,' runs an anonymous Irish poem from the ninth century, 'to be without a soft bed; a cold frosty dwelling, harshness of snowy wind.' While another nameless poet of the same period revels in the spring: 'The harp of the wood plays melody, its music brings perfect peace, colour has settled on every hill, haze on the lake of full water.'

I walked on around the coast that day. At dawn it was dull and cold but within a few hours sunlight filled the morning, brighter than it had been all year, covering the cliffs and little bays in dusty yellow light. Near the Rumps I found a place in the rocks that was out of the wind, and where the sea shone below and tiny buds of stonecrop bordered the bare rock. It was less warmth itself than an absence of cold that made me dawdle there, and dip into the epics of Geoffrey: 'A certain giant of marvellous bigness arrived out of the parts of Spain . . .'

I listened to the lick of surf on the rocks. I heard the whine of a small aircraft . . . the companionable song of a stonechat . . . rooks . . . When I woke, it was with that blank after-sleep feeling of deep comfort. The sea below was blinding. Squinting down into it, I watched it sparkle and saw the breezes spreading over it, and thought of Gildas: 'Waters glide with gentle murmur, guaranteeing sweet sleep for those who lie upon their banks.'

Back at Ardevora, there were no gliding waters – no water at all in fact. The old pump had broken weeks ago and the new one had not yet arrived. We were still renting the cottage next door; with its own borehole and underfloor heating, it was becoming hard to leave. I had arrived back from the north coast after dark, and first thing next morning crunched up the frosty track to the house. The sun was just up, a red glow rising through bare branches. Every surface, every length of scaffolding board, every discarded brick and broken tile was covered with a fur of tiny ice crystals. The mound of rubble in front of the house had grown and was now frozen solid again – bristling with old pipes and broken buckets and yellow Mastercrete bags.

That evening, after dark, the new pump arrived. There had been a good deal of debate about whether to continue with a well-head pump or go with a more expensive submersible. A law of flow mechanics, I was told, means that the limit of sucking water up from above, as opposed to pushing it from below, was somewhere between nine and twelve metres. Our well was just over ten metres deep. We had opted for prudence and that evening the new submersible pump was brought by the ebullient father-and-son team who were to install it. Beside the well, with its knee-high brushed steel column, its curl of green pipe leading up to a large black bulb, the pump looked like a bizarre musical instrument. I took photographs. It was the last time I'd see it and we would depend upon its smooth functioning now for years.

Inch by inch we lowered the device into the well, paying out its electric cable, the blue feed-pipe and a length of string as a hoist. Its shape grew fainter in the shaft. The glare of our head torches shone on the water at the bottom. A piece of grit fell, and with a loud *plo-op* wrinkled the deep-down mercury surface. The pump dropped on past the walls. I'd always imagined that although the well was now built into the house, its meticulous bore of stone-work and chiselled rock was much older, that buckets had been lowered and raised from it for hundreds of years. Looking at its walls, I felt once again that sense of awe at the thought of those who'd been here in the past, who'd dug the well, hacking down through the bedrock with only hand tools and a dowser's word that there'd be water. But I also found it faintly disturbing to see quite how far down the water was.

Last summer there had been the loss of power for the pump. There'd also been, after a rainless six weeks, a time when the well had run dry. I remembered turning and turning the faucet outside and the shock of nothing happening. I had been reading of the medieval settlements on Bodmin Moor, abandoned when the climate became drier, and I pictured the ancient grassy ruins up there, and projected their equivalent down here. It turned out not that the water table had dropped, but that the seepage rate could not replenish the well as fast as we were drawing it. We had waited and it did come back. But the fear remained.

The pump reached the water – but the pipe had a kink in it. We would have to bring it back up. Then there was a snag with the filters. It was late in the evening before the inspection plate was finally screwed back on. We were ready. I watched the pump drop again, sliding into the void, trailing its tethers of pipe and cable. The electrics were flicked on. The switch in the pressure-vessel tripped, and there was a faint hissing as the water rose from below, flooded through the pipes and up into the holding tanks in the roof.

We stood smiling with relief, sharing the first glass of water and listening to the pipes, and in that flowing sound I sensed life returning to the body of the house.

# 7 | GLASTONBURY

Anglo-Saxon *Glaestingabyrig*: 'town of Glaesting'; *glaston-*, possibly from Celtic *glast-*, 'blue-green' or 'woad'.

St dunstan had a dream. He saw a tree, and the tree's branches stretched out far beyond the horizon and in those branches he saw hanging the cowls of monks, and in the crown of the tree he saw one large cowl, which belonged to the great monk Aethelwold. The tree was so tall that it covered all of England with its shade, and Dunstan went to a priest with hair as white as an angel and the priest explained the dream: *monks and monasteries will be the protection of the nation until Judgment Day*. In the dream St Dunstan had seen that the tree, with its deep roots and its wide trunk, rose from the monastic precinct of Glastonbury.

Glastonbury is the mythic heart of England. It's where Christianity arrived, where the first church was built, where King Arthur was buried. Few places can claim such a record of mystical experience, nor such a range of wonders and attractions over so long a time. Like Britain, Glastonbury is also a *locus amoenus,* described in *The Chronicle of Glastonbury Abbey* as 'spread wide with numerous inlets, surrounded by lakes and full of fish and rivers, suitable for human use, and, what is more important, endowed by God with sacred gifts'. Glastonbury housed, until the Dissolution, the greatest collection of relics in the country, the greatest stories. It was a one-stop pilgrimage site, reducing the idea of earthly sanctity and of the nation to a single place.

Untangle all the tales, the dreams, the visions and theories and there's little to explain them other than the location itself – the sandstone cap of the tor and the ruined tower of St Michael's upon it. Iron in the water has seeped into the rock and oxidised, so that in parts, rather than eroding over the years, the tor has actually been hardening. Spend any time on the Somerset Levels or on the southern slope of the Mendips and this small hill haunts you with its constant presence. As a child growing up nearby, I found it was always *there,* with its pudding shape and the tower of St Michael's on top like a single candle. Few places can match its striking visibility, given that it's not that big. Like some clumsy spook, it peeps over ridges, loiters in the gap between features which ought to hide it. When you can't see it, you find yourself looking again, and nine times out of ten you'll spot it.

In 1960, shortly before I was born, my parents bought their own run-down farmhouse on the northern edge of the Mendips. My mother spent her pregnancy painting the walls, and has always joked about what the paint fumes must have done for my foetal development. When they moved in, I was a few months old. A ladder led to the upper floor and water came from a well outside. Year by year the house filled with the technological advances of the age – mains water, mains electricity, an electric cooker, central heating, freezers and fridges, a Sodastream and stereo, a colour television, an answering machine, a flat-screen, lap-tops and broadband wi-fi.

Now my parents were in their eighties, and they were moving out. They wanted something smaller, 'more manageable'. To me, trying to establish our own house, aware of just how much we had taken on, their decision made sense. But when we went to see them in late February, and pulled into the village square and saw the estate agent's board in the garden, I had a sudden feeling of emptiness. I had spread my wings far beyond this village, beyond the hill and the combe. But this house had always been here, and it had always been home.

That evening, we talked for the first time in detail about how they'd found it. They were both about thirty, renting a flat in Bristol.

My father had just started work as an architect in the city. What he wanted was an old house to restore, and each weekend they toured the Somerset villages in a Morris Minor, my elder brother lying in a cot on the back seat. One dreary day in December they arrived here, at the foot of the Mendip Hills. They parked in the square. They saw the school building and the church behind it. Beside the school and the church was an old farmhouse – and a *For Sale* board. Inside, at one end of the house, a few cattle were wintering; at the other end, in the kitchen, lay Mr Riddel on a broken sofa: 'all bristly and drunk', recalled my mother. A small window above Mr Riddel was the only source of daylight.

My mother said she knew, even as she got out of the car in the square, that this was where she wanted to be. My father was already thinking ground plans, load-bearing walls, bigger windows, pre-stressed lintels. They scraped together a couple of thousand pounds and bought it within a week. They'd now been here for over half a century.

Early the next morning, I rose at dawn to walk over the hill to Glastonbury. I'd tried once before, one January years earlier, but fell in a rhyne down on the Levels and lost heart. Now, before my parents left, I wanted to try again. I let myself out through the garden door, trotted up the steps to the lawn, past the old Elsan and over the fence. The footpath ran from the church across the glebe and into the combe. I passed the Rock of Ages, and Aveline's Hole with its slanting gape and stub-tooth rock in the entrance. A French lorry stood in a lay-by near it, dwarfed by the cliff above; I saw the just-woken driver climb down, stretching and yawning as he lit a cigarette. Leaving the combe, I entered the steep-sided valley that used to lead us to Goatchurch Cavern.

The stream at the bottom was February-full and I splashed up it. At the spring a glassy arc of water poured from a pool in an old cistern. The bottom was just visible, spotted grey with pebbles. For a moment I stood there transfixed by the clarity of the water and the illusion it gave of perpetual motion, gushing out at my feet but with no visible sign of where it came in.

Up over the top to Charterhouse and Velvet Bottom. The clouds were black and heavy above Cheddar. By the time I reached Priddy rain was driving in hard on a cold south-easterly. I walked on, head down, watching the tea-brown streams running back past my splashing boots. When the streams thinned and stopped, and started to flow with me, I glanced up. There was nothing visible yet beyond the Mendip ridge. I carried on. I must have looked up a dozen times until – with the slope now dropping steeply towards Ebbor Gorge – I saw the Levels beyond Wookey stretching out into the distance. And there, like an old friend, the Tor. In the rain, it played a new trick. From where I stood, the distant skyline of the Blackdown Hills bisected the summit, so that it looked as if St Michael's tower stood on both peaks, or neither. Coming down the hill, the rain eased. I watched the low clouds part and let through not the sun's full rays but a diffuse yellow light, as if a torch had been shone down through a trap-door, sweeping across the fields until it found the Tor.

St Dunstan's dream of Glastonbury was part of a long tradition of revelations. When St Collen travelled to the court of Gwyn, King of the Fairies and leader of the Wild Hunt, he slipped out of this world via the Tor. Beside Glastonbury Abbey, the site of the lost chapel of Edgar was discovered in the early twentieth century during a seance, its plan accurately sketched in a bout of automatic writing. In the 1930s, using maps and aerial photographs, the artist Katharine Maltwood spotted in the paths and boundaries around Glastonbury a series of deliberately arranged Zodiac figures, all centred on the Tor itself. Dowsing their way up from Cornwall, Paul Broadhurst and Hamish Miller followed the Mary and Michael 'energy lines' to Glastonbury; at the Tor, the rods picked up an elaborate interlocking pattern for the two lines which, when drawn on paper, revealed a diagrammatic image of sexual union. In his 1932 novel *A Glastonbury Romance* John Cowper Powys wrote: 'There are half a dozen reservoirs of world-magic on the whole surface of the globe – Jerusalem . . . Rome . . . Mecca . . . Lhasa . . . – and of these Glastonbury has the largest residue of unused power.' When John Steinbeck first saw the Tor from Cadbury Castle in 1959, he wept. He spent nine months in a rented cottage nearby and one night discovered he could see the Tor through a gap in the hedge. 'Am I in any way getting over to you the sense of wonder,' he wrote to a friend, 'the almost breathless thing?'

Four centuries earlier, in June 1533, Henry VIII's antiquary, John Leland, was also left breathless by the sight of Glastonbury. It was a moment that would not only set the course of his life's work but initiate the entire subject of British topography.

It had been an auspicious summer, full of hope for both Leland and the country. A few weeks earlier he had attended the triumphant entry into London of Anne Boleyn; his own verses had been read out during the ceremonies. Now the King had despatched him 'to peruse and dylygentlye to searche' every library, monastery and college for manuscripts that might reveal more of the nation's glorious antiquity. En route to Glastonbury, Leland had conducted bouts of intense study in a string of monastic libraries – from Guildford to Christchurch, Sherborne to Plympton. Now, riding up from Taunton, he came to the 'most ancient and at the same time the most famous monastery in the whole island'. Seeing the Tor rising from the Levels had struck Leland so deeply that he arrived at the abbey already in a state of elation. He had planned to rest, to put off his work until the next day, 'to refresh my mind'. But he became aware of a strange urgency: 'a burning desire to read and learn inflamed me'. He hurried through the cloisters of the quad to the library. 'Scarcely had I crossed the threshold when the mere sight of the most ancient books took my mind . . . I stopped in my tracks.' Standing in the doorway, gazing at the books inside, he found himself overcome by 'an awe or stupor of some kind'.

The sight propelled John Leland on a quest to catalogue the country's wonders, first in monastic libraries, then in the land itself. For twelve years, he travelled around England, recording the byways and villages, the chartered towns and valleys, the harbours and battlefields, all the nation's sites of great deeds. He wrote to Henry saying he was now 'totally inflamed with a love to see thoroughly all those partes of this your opulente and ample realme'. He would miss neither 'cape, nor bay, haven, creke or peere, river or confluence of rivers, breeches, waschis, lakes, meres, fenny waters, montaynes, valleis, mores, hethes, forestes, woodes'. In a plain and oddly poignant phrase, he later told the King: 'I have seen them, and noted in so doing a whole world of things very memorable.'

The figure of the roving antiquary was not new. In the fifteenth century John Rous, William Worcester and John Hardyng had all travelled widely in pursuit of England's past. But until Leland no one had shown an interest in places for their own sake, nor ridden so far to see them. William Camden drew on Leland extensively when, in the formative decades of Elizabeth I's reign, he wrote his *Britannia*, perhaps the most influential of all national topographies. For the next two hundred years Leland's work was used by almost every historian and topographer of Britain; it is surprising how many local histories still begin with one of his details.

During his lifetime, Leland never managed to draw all his researches together; much remained in note form. Within decades of his death, they had begun to decay. Damp seeped up from the foot of the pages. Several books were lost, sheets were scattered. By the

1570s the desk-bound topographer William Harrison had trouble using them (it was Leland's fate that his notes were plundered and plagiarised). The folios, complained Harrison, were 'moth-eaten, mouldy and rotten... falling to pieces every day'. What survives of them is now bound into eight volumes in the British Library, an incomplete monument to the past to add to all the other traces of sacred sites, the ruined monasteries, the gapped stone circles.

John Leland is an intriguing figure. Janus-like, he looks both backwards into the Middle Ages, and forwards to the Enlightenment. His notes are full of the flavour and detail of the medieval world, yet he was the first to apply an empirical method to recording place. He believed firmly in Geoffrey's British History and scorned its sceptics – 'a plague on those dunces' – and yet, as a student in Paris in the 1520s, he was so caught up in the excitement of humanism that he has been called a 'bridge... which connects England with the Continental Renaissance'.

He travelled alone, riding from town to town, curious about everything and everywhere. No one had ever roamed and recorded the country with John Leland's open-eyed attention. No one since has seen what he saw. In his work, the land comes alive, in all its diversity; the world grows larger. He took down details of those who owned estates, of their hereditíes and great houses, of ancient castles and churches. He examined charters and recounted historical incidents, local legends and lore. He addressed the hour-by-hour concerns of wayfarers like himself – the precise distances between places, the course of rivers and their crossing-points, hills and passes.

In the sixteenth century, people were still drawing *Orbis-terrara*, the schematic maps that placed Jerusalem at the centre of the world and separated the continents with straight lines. Topography was more abstract than literal, more theocentric than cartographic. Beginning at Glastonbury in 1533, John Leland's approach represented something new. Gone was the medieval polarity of hostile wilderness versus the *locus amoenus*. Instead, Leland wrote what he saw and what he heard, invoking the land to create a fresh sense of the national past – one based on the soil, on hills and valleys and rivers and streams and inlets, on the experience of those who lived in it. Behind his footslog and note-taking lay a driving and unshakeable love. His passion for ancient manuscripts and for the land itself were two sides of the same coin: *this is us, this is Britain*.

By the time I reached Glastonbury's High Street, footsore and mud-spattered, it was mid-afternoon. Dropping down from the Mendips, I'd watched blue sky spread from over the Severn Estuary. Now the sun was everywhere, glinting from the roofs of cars on the Wells road, bright gold on the slopes of the Tor and on the standing-stone shape of St Michael's.

I was almost too late for the Abbey. The gatekeeper looked at his watch. 'The monastery always welcomed pilgrims – like yourself.' He took in my walk-stained trousers. 'Half an hour only!'

It had been years since I'd been here and I'd forgotten this: amidst the crocuses and the grassy banks and the edged pathways, the *loss*. The Abbey walls glowed in the late sun, all truncated arches and snapped-off vaults, like a stage direction that calls for 'vanished glory'. 'O Glastonbury, Glastonbury,' wailed John Dee, thirty years after the Dissolution. 'How Lamentable is thy case, now? How hath Hypocrisie and Pride wrought thy Desolation?' While Michael Drayton asked: 'What? Did so many kings do honour to that place, for Avarice at last so vilely to deface?'

It is hard now to imagine the full scale of Glastonbury before the Dissolution – not just the Abbey and the cloisters, but the combination of relics, artefacts and stories that made it the country's most popular place of pilgrimage. After long tramping through the byways of southern England, pilgrims could whisper their devotions before a piece of Moses's Rod, drops of milk from the Virgin Mary, a section of the Crown of Thorns, the teeth of the Apostles. They could gaze at the glittering feretories, the bone-casks of St Dunstan, St Patrick, King Edgar and St Gildas, visit the cemetery in which lay the remains of old King Coel (grandfather of the Byzantine Emperor Constantine), and see the pyramids beneath which had been discovered the body of King Arthur. For those who could read, there was a copy of the story of how Joseph of Arimathea built a church to his niece here, St Mary the Blessed Virgin – the first ever church. Glastonbury, wrote William of Malmesbury in the twelfth century, was a 'heavenly sanctuary on earth'.

In the summer of 1533 John Leland was shown all this by Richard Whiting, Glastonbury's kindly old abbot. Leland saw the black marble mausoleum containing the bodies of Arthur and Guinevere; he held and measured the cross that was found with Arthur's remains. Within a decade of Leland's visit, Glastonbury's treasures were scattered, the monastery destroyed. Richard Whiting, then in his eighties, was dragged through the town strapped to a hurdle. He was taken to the top of the Tor and hanged. His head was cut off and placed on a spike outside the Abbey, his body sliced into four and the pieces displayed around Somerset.

The savagery at Glastonbury was motivated less by avarice than by royal pique. It was Herod massacring the innocents, fearful of the threat they might make to his crown. Many at the time believed Glastonbury to be the second Rome and Henry did to it what he would have liked to have done to Rome itself. Its buildings were abandoned and with them went their cache of lore, relics and associations, the mythical reputation that had spread beyond these shores and had been accumulating since long before the Tudors, the Plantagenets and the Normans, before St Dunstan and the Saxons, before Joseph of Arimathea, back before the Lake Villages of Meare and Godney and the Neolithic earthworks on the Tor, ever since the first settlers saw the odd formation of the hill rising from the water.

Henry nearly succeeded. For centuries Glastonbury stood as a ruin, visited only by a few curious antiquaries. But during the last century or so, its appeal has been growing – no longer as a place of Christian pilgrimage, nor as a site with the Arthurian emphasis of Tintagel, but as a combination of these and more, a post-modern stew of myths and traditions. Maybe it really is the energy lines, those hidden channels of power that converge at this place to give it its aura and generate its miracles. Or maybe it's simply the shape of the hill.

The Abbey was closing. Finding myself back out on the street, I tucked into a busy courtyard named 'The Glastonbury Experience'. Here was the Brigit Healing Centre and a shop called The Stone Age, which sold crystals and offered in an upstairs room sessions in Crystal Sound Activation (*Reawaken your inner fire – Return to your Source Frequency…*). Outside the healing centre, I picked up a list of the town's February events and was dizzied further by offers of a Shamanic Dance Trance, Bellydance Burlesque classes, Barefoot Boogie and Bringing Your Soul Home.

Like Tintagel, Glastonbury offers everything: Arthuriana, antiquity, crystals, a multitude of paths to a better life. I thought of John Leland, also awed by the range of what was on offer in the Abbey library. He made two discoveries – the *Charter of St Patrick* and Geoffrey of Monmouth's *Life of Merlin* – and examined countless other manuscripts. Over the next few years, in monastic libraries elsewhere, Leland managed to study a huge number of texts, until the forces of the Dissolution began to scatter them. It wasn't long before his friend John Bale was reporting that vellum manuscripts were being sold to soap-sellers for wrapping, and loose folios were falling around cloisters like autumn leaves. They were used to clean candlesticks, to polish boots. John Leland wrote to Thomas Cromwell urging him at least to preserve the libraries and their 'noble antiquities'. But it did no good.

Oddly the destruction of the monasteries did nothing to dent Leland's patriotism, nor his devotion to the King. In the mid-1540s, he started to gather together his papers and notes. He was quite convinced that, presented correctly, they would be the greatest ever representation of the realm as a celebrated place, its earthly splendour and its past. He would assemble a full account of the antiquities of the King's dominions in England and Wales under the title *De Antiquitate Britannica* – a volume for each province, say about fifty in all. Another six volumes would cover the islands. In *De Viris Illustribus*, Leland would present the worthiest, the most heroic of the nation's past lives – in four volumes. Another three would detail the royal houses of Britain. Leland assured the King that, by the publishing of such works, his greatness would never be forgotten. All of it, stressed Leland, every phrase and every paragraph, would be written in 'a flourishing style' to ensure its appeal for generations to come. With the books, the King would also receive the *Liber de topographia britanniae*, a table made of silver. Etched into it would be a map detailing the marvellous extent of his kingdom.

By 1547, when Henry died, Leland had been unable to present him with any of his intended offerings. Soon afterwards, he suffered a breakdown. One scholar has diagnosed a 'magpie' complex; others believe that it was the sheer intensity of his work. A recent analysis identified him as bi-polar. Leland's own descriptions offer hints of hyper-sensitivity – he had stood in an awestruck 'stupor' at the door of Glastonbury's library, and was 'inflamed' with the desire to learn, and 'totally inflamed' again with the thought of exploring the land.

Leland had burnt himself out. What he saw of the land, of the soon-to-be-destroyed monastery libraries, engulfed him. He spent his last years in the care of his elder brother, incapable of work. He died in 1552.

Some months after walking to Glastonbury, I visited my parents' house in Somerset for the last time. It had been sold; they had already moved out. Within days the new owners would arrive. All the furniture had gone, the curtains, the rugs, the kitchen table, the photographs. What was left were hollows and imprints – the empty bookshelves and unfilled alcoves, the darker squares of paintwork where pictures had hung, the little pressed-down patches of carpet beneath the armchairs, the floorboards pocked with high-heel marks from a house-warming dance they gave nearly fifty years ago.

There were several details which, in all my years there, I had never fully examined – the full scale of the bevelled beam in the kitchen, my father's star-design on the newel post, the beautiful shape of the brass door handles, and the corbels of the stained-oak mantelpiece in the sitting room, now stripped of its figurines and brass carriage-clock. All the features that gave the house its charm, old fireplaces and not-straight walls, now looked cold. The memories of the place had separated from the place itself. It was as lifeless as a corpse.

Upstairs, I stood in my parents' old bedroom, a large-windowed space which looked out over the edge of the garden, over the churchyard with its skewed headstones, into the glebe and towards the combe. Mendip Wood already showed a faint brushing of new-growth green; behind the church was the rise towards Long Rock and the ferny expanse that stretched over the skyline to Cheddar. Only then, looking at the familiar curves of the land, did a tide of nostalgia sweep over me. I saw my mother's daffodils beneath the magnolia and the low wall behind and was struck suddenly by the effort of it all, the years spent restoring the outbuildings and the garden, and now everything gone, and all of us living elsewhere.

Towards the end of February, in Cornwall, we moved back into our own house. The main building work was finished and while the project was nowhere near completion – no carpets, no curtains, bare drying plaster, whole rooms inaccessible – it was habitable. I kept the stoves burning all day, feeding them with the trees we'd felled last year. I'd been chopping and cleaving and stacking, learning the craft of the woodman, with its rewards (smell of split wood, the beauty of the grain) and its novice pitfalls (aching shoulders, bruised thumb and constantly jammed axe).

Late at night as the stoves dimmed and cooled, the iron casing would make a clicking sound. Lying in bed later, I listened to the house making a similar noise, creaking like an old ship. I imagined it settling down again, readjusting to the lie of its new timbers, getting used to its refit. And from the creek came the sound that now spelt 'home', the precise note of the curlew's cry.

# PART II

# 8 | HENSBARROW

'Hens-' from English 'hind', '-barrow' from Old English *beorg*, 'tumulus, grave-mound'. A nearby farm, Cocksbarrow (now swallowed by Littlejohn's China Clayworks), 'must be named as a joke, based on the modern form' (Padel).

On the track that leads down to the house is a bend, and as you come around it, you can see for the first time the full expanse of Ruan Creek. If the tide is in, the creek looks like a large lake, edged by salt marsh and framed by oak-covered slopes. At low water, the mud banks extend for half a mile or so – a glassy surface still wet from the ebb, not quite sea and not quite land. But at neaps, several days can pass without the flats receiving any water at all and at those times the sun has time to dry out the mud and bake it until it turns a very particular colour: a pure pale ochre, with a curious opacity. The old river valley is filled, in places to a depth of fifty metres, with china-clay spoil.

It is hard to think of the creeks here affected by industry. They look untouched – trees lie in the mud where they fall; the only sound comes from geese and waders. Yet without the silting, the valley would never have become so isolated, would now have many more people living and working in it, and many fewer birds. The spoil is no longer 'refuse' but 'habitat', full of crustacea and worms and protected by law as an SSSI ('site of special scientific interest').

As you round the bend, there also appears a much more obvious sign of the china-clay industry. Along the skyline, some fifteen miles to the north-east, runs a line of peaks – higher than anywhere in the whole of central Cornwall, and made from the same whitish mica-sand that clogs the creek. Walking down the track, as I did now each day, spotting the distant heads of the spoil-heaps and the silted-up creek below, always made me think of the way dogs run up and look at you with an expression that is part pride and part guilt, as they dump something dead at your feet.

In early March, I took a bus back to Wadebridge and resumed the westward hike towards Land's End, with enough free time ahead to follow a route across the spine of Cornwall, through the china-clay area and on down the Fal. The ancient and medieval beliefs that lay over Bodmin Moor, Tintagel and Glastonbury were behind me; ahead lay places bearing the rather more physical imprints that followed the Enlightenment.

Near Withiel I climbed out of the valley and up a steep-sided lane. The sun shone bright on the ribbon of tarmac; in the hedgerow were the first celandines. Fresh alexanders had pushed their green stalks through the tangle of last year's growth and I leaned in close to look for spring. There! Tiny stars of inflorescent buds – the sappy, vegetable smell.

At the top of the hill, the view opened out and the clay spoil-heaps appeared again – larger and darker from this aspect, from the north. Their slopes were in shadow but I could make out a faint gossamer of high-voltage cables and the dots of village sprawl and the faint snow-coloured areas of recent dumping.

The town of Roche is named after a lump of grey quartz and black tourmaline which rises like an outsize dolmen from a nearby piece of rough ground. Clambering up a rusty metal ladder, I came out on top of the rock. As at Glastonbury, the summit featured a roofless shrine to St Michael (it lies on the dowsers' 'Michael line'). And this was mineral country again, one of those places in Cornwall where you sense the fragility of the earth's crust. In the exposed rocks and the shape of the hills are signs of that ancient rupture, the moment when boiling magma thrust its way through layers of benign sediments, its superheated gases hissing along fissures, transforming them into a hundred crystallised compounds. Of all the world's minerals, fifteen per cent can be found in Cornwall – and almost all of them around the granite lumps that run down the peninsula like stepping stones – Bodmin Moor, Carnmenellis, West Penwith, the Isles of Scilly – and here around Hensbarrow Beacon.

We think of granite as hard, sturdy enough for lighthouses and kerbs, gravestones and mooring-blocks. But in places, during cooling, gases degraded its feldspar so completely that a jet of water can now easily reduce the rock to a white soup. Rinse the mica and quartz from the soup and you have kaolinite or china clay. According to the industry's historian, around Hensbarrow are 'the finest resources of china-stone and china-clay in the world'. An area once noted only for rough pasture and small-scale tin-works is now known simply as Clay Country.

The path pushed on through a scrappy landscape. There were a lot of bungalows. Big dogs barked from small homes. Cornish flags fluttered from high poles. In a yard full of wrecked cars was a Transit van with no wheels and on the side: *Kandy's of Exeter, Beach goods, Gifts etc.* Beside it was a mobile home from which stepped a woman in peignoir and overcoat. She bent to water a tub of daffodils and gave me a glance long enough to remain with me for the rest of the day. Such are the hazards of the lone walker.

For thousands of years, Hensbarrow Beacon was one of Cornwall's most prominent peaks. It's there in the name – a Neolithic hill-top barrow from which both Rough Tor and Carn Brea were visible, enlarged in the Middle Ages to form a beacon. On top of it, I found the mini-obelisk of a trig point, with its patinated socket for a theodolite. Visibility was always Hensbarrow's gift to the region. But not any more. Neither Rough Tor nor Carn Brea can be seen now and there can be few trig points in the country from where, just a hundred metres away, the ground rises. You have to tilt your head to take in the full size of the man-made mountain beside it – the spoil-heap of Gunheath China Clay works.

I scrambled up its lower sides, and reached the service track. The whole china-clay area is prohibitive with its fear-inducing signage – THIS SITE IS NOT A PLAY AREA...WARNING...BLASTING... DANGER – and the constant thump and clank of heavy machinery, But no one was about so I carried on, up through a series of switch-back bends. Everything was white. The compacted surface was white, the puddles on it were white and the small sprigs of heather and furze that had tenanted the white slopes were brushed with a fine white layer. The Cornish poet Charles Causley wrote of this area, a little cryptically: 'Here is the lunar and lunatic landscape of the moon: a weird white world dusted over with the colour of sex.'

According to my Ordnance Survey Explorer, I was standing nowhere. Just to the north, the map was speckled, signifying 'Land-fill or slag/spoil-heap'. But here it was simply blank. Contours do not enter the area. They dangle. If the whiteness looks pretty much like the landscape itself, it also leaves an impression of furtiveness, as if it was a closed-off military zone, the testing ground for some fiendish weapon. But even military zones have contours. Around Gunheath were three or four other giant pits and together they make up some six square miles of cartographic void – a graphic example of how the hunger for commodities has purged areas of their singularity, of the triumph of 'space' over 'place'.

I pressed on, waiting for someone to tell me to stop. The sides of the road were crumbling away. Rainwater channels had sliced gullies down into them, leaving lone pinnacles hatted with single rocks, and running down to join the road below in mini-deltas of fine white sand. This is how land looks when it has just formed. The road pushed higher – then all at once I was out on top, standing in a wide-open plateau of glacial pallor, beneath a cloudless sky. St Austell Bay glittered far-off to the south. To the north the sea spread its steely sheen out beyond Newquay; below it, traffic on the A30 ran like water down a fishing line; Rough Tor and Brown Willy stood on the distant horizon. It took a fair bit of binocular-scanning but in the end, beyond the bluffs of the Fal Valley to the south, in amongst the patchwork of fields and darker woodland, I made out the faint smudge where the barn on our track should be and the way leading down to our house.

Crouching to keep my figure less visible, I approached the edge of the pit. I lay down. The mud was creamy beneath my palms. I shuffled forward between two white boulders, and looked down. For a moment, I couldn't focus. The scale was dizzying. I could see a giant marble-run of zigzagging ledges and on one a tiny shovel-loader scooping the fall from a recent blast, filling a matchbox-size truck. It was baffling to think that such vehicles, and before them horse-drawn carts, could have created such a hole.

Right at the bottom of the pit, below all the icing-white terraces, was a pool of water. The wash had collected in a hollow from which it could seep no lower. I spent a long time watching the trucks work up and down the side of the pit, pausing by the digger to have their backs filled with soft white rock, creeping down to dump their cargo, before labouring up again. But my gaze kept returning to that watery hollow. Its colour lay somewhere between turquoise and cobalt and it filled that entire white hole, scoured from the side of the earth, with its chill and heart-stopping beauty.

In the Cornwall Record Office, the land documents for this area – until fairly late in the eighteenth century – do not mention china clay. There are leases for small tin-mines, licences for tinners to run leats over so-and-so's property, agreements by surgeons and inn-keepers to set up tin ventures with names like *Happy Chance, Fair and Honest* and *Take-more-care-next.* But it was 1770 before the records show any sign of the substance that would sweep aside every one of these small-scale operations. The document offers one William Cookworthy 'liberty, license and authority to search for and raise such materials as china-clay and china-stone'.

William Cookworthy was neither an industrialist nor a speculator. He was a Quaker and a chemist – 'one of the greatest chymists this country has ever produced'. He was also the father of the china-clay industry. But he would have been horrified by the vast pits that his discoveries spawned. For him the land beneath his feet was a storehouse of elements which, if extracted, separated, re-mixed, and heated in the right way, could perform extraordinary tricks, make blades, give strength to buildings and cure diseases. The careful separating of God's materials was a pursuit which, like prayer, was an end in itself, a process which allowed contemplation of Creation, which might prove of some direct use but whose primary benefit was to occupy mind and body in a way which nourished the soul.

Cookworthy combined the secular and the religious, the scientific and the mystical in a way that was impossible even a few decades after his death. He had a chemical practice in Plymouth but spent much of his time riding out into the country, on the West Country circuit of the Society of Friends, which allowed him to search mineral-rich Cornwall for materials.

For many years his mineral forays had been driven by one particular project: the quest for true porcelain. It was a puzzle that had exercised Europe's great minds ever since Marco Polo's return from China in the thirteenth century. Porcelain is not like other pottery in which the clay and glaze are separate. 'China stone', or 'petuntse', is used for the body and china clay for the glaze, and at very high temperatures they perform a miraculous fusion.

For two decades Cookworthy had been experimenting with materials from further west in Cornwall. But at a place called Carloggas (the name means 'rock pile of mice') he took samples of the

Hensbarrow clay. Back in Plymouth he blended it with petuntse found near the clay. He mixed them, he ground them, he fired them. He mixed them again. He tried varying proportions, in different kilns, fired with different fuels at different temperatures, for different lengths of time. Years passed. Then one day in 1768, he pulled a sample from the kiln, tapped it, examined its vitrification and realised that he had produced what no one else ever had in Britain – true hard-paste porcelain. Soon afterwards, HM Patent Office issued patent no. 1096: 'Manufacture of Porcelain – Cookworthy's Specification'.

The appeal of porcelain lay in a combination of its rarity and its magical properties. Pieces from China had been trickling into Europe for hundreds of years, but the land route along the Silk Road was too bumpy for transporting china in any quantity. It wasn't until the Portuguese opened up a sea route around the Cape of Good Hope that Chinese porcelain became more widely available – which only served to increase demand. By the beginning of the eighteenth century, porcelain was more valuable than gold.

It is unlike any other known material. Light can seep through it. It has the delicacy of lace but is astonishingly robust (steel cannot cut it). It can withstand very high temperatures and when tapped gives off not the dull click of pottery but a chime. Drink from a porcelain cup, it is said, and you'll become immune to poison. Porcelain became the favoured gift of kings to kings. In Mantegna's *Adoration of the Magi*, painted in 1462, the exotic offerings are carried in what look like porcelain containers. Bellini's 1514 *Feast of the Gods* shows the food served on porcelain plates. John Dryden described 'porcelain clay' as 'the workmanship of heaven'. The imported pieces were astonishing shapes – all twisting dragons and jewels embedded in the handles. They brought with them something more than just their substance. In their whorled voids and glistening curves was contained the myth of far-off Cathay.

Something similar had happened with certain objects in the Neolithic. Archaeologists became curious when they found that greenstone axe-heads had been deposited in rivers unused. The sources of those axes turned out to be very few in number. More than a quarter of those discovered in England were chiselled from one particular rock face at Pike O'Stickle in Cumbria. They did not come from the foot of the cliff, where the same stone occurs within easy reach, but from high up, hammered from the lode while balancing on a narrow ledge. The axes, it appears, were not weapons or tools so much as prestige goods, exchanged with regard for their dramatic and dangerous provenance. Other such ceremonial axes have been traced to the far west of Cornwall, from the distinctive headlands, the 'cliff-castles', now thought to have been used themselves as ritual sites. Likewise the clays of Cornwall's remote Lizard peninsula – an area whose geology is so unusual it is matched only at Mount Olympus – produced pots in prehistory that ended up in sites all over southern England. Axe-heads, pottery and porcelain retain the memory of their source, the story of their extraction. They were narratives in physical form; they were places, carried.

The Chinese had helped sustain the mystique of porcelain – and its value – by keeping its manufacture secret. In Europe, discovering the materials and the processes became a quest every bit as seductive as that of alchemists' search for the philosopher's stone (often it was carried out by the same people). Did the Chinese, as Marco Polo reported, dig up their china clay and leave it to season in mounds for forty years? Was it in fact gypsum buried for a century, or common clay mixed with some arcane combination of glass, eggshell and the ground-up carapace of a lobster?

In Venice and Florence, Rouen and St Cloud, at Chelsea and Bow, Europeans produced porcelain – but none of it was true porcelain. Then at Meissen, in 1709, among the torture chambers of

King Augustus of Saxony, the alchemist Johann Friedrich Böttger achieved the mystical fusion of glaze and slip but, lest he reveal the secret, was kept under close guard for the rest of his life.

At about the same time, a letter from a Jesuit priest in China began to circulate in Europe. Reaching the closed region where porcelain was made, Père d'Entrecolles was amazed to find Jingdezhen – a city of a million souls – given over to the manufacture of one product. Blind men squatted in alleys grinding the pigments. Three thousand kilns burned continuously. From across the plain, he reported, the city glowed like 'one large multi-vented furnace'.

D'Entrecolles' account set dozens of manufacturers all over Europe experimenting. Yet for all the details he gave, the secret proved just as elusive. In Britain, it was more than half a century before William Cookworthy managed to combine the right materials in just the right way. During his arduous years of experimentation, he kept his patron informed through letters. Thomas Pitt owned the land at Carloggas and maintained a stake in Cookworthy's project. The letters survive among the Fortescue Papers in the Cornwall Record Office, each one sealed with blood-red wax and addressed: *Thomas Pitt Esq., His House near Piccadilly, London.* 'Esteemed friend . . .' they begin. Reading them now, it is possible to trace, even in Cookworthy's measured tones, the arc of excitement, experiment – and triumph.

They tell of a tense and competitive period. Someone comes snooping around Carloggas – 'a bold fellow of the Projecting kind' – offering money to be shown where to find more china clay. With their patent still pending, Cookworthy and Pitt risk losing their advantage. Fortunately Pitt's nephew is the Prime Minister and, once the problem is highlighted, the patent is swiftly issued. With the technique mastered, and using the high-grade materials from Carloggas, Cookworthy began production – first at his pottery in

Plymouth and then in Bristol. He engaged a partner and dozens of craftsmen. They made sauceboats and teapots and cider mugs, figurines of the Four Seasons, vases and dessert baskets. A good deal of the pieces went to America. But none of it made them any money. The delicate balance of material, of flux (the right quantity of lime and fern ashes), the right kiln (wood-fired, thirty-six-hole), was hard to maintain on a large scale.

Cookworthy died in 1780, his partner went bankrupt, the patent ran out and commercial men with names like Minton, Wedgwood and Spode bought up the mineral rights around Carloggas. They transported the china clay back to their potteries in the Midlands. Like Cookworthy, they found it hard to manufacture hard-paste porcelain on a commercial scale, so they mixed it with fifty per cent bone. Only Cookworthy's potteries ever produced true porcelain in Britain.

Demand for the clay continued to grow. Apart from bone china, it was used to whiten paper, to bleach cotton, and as a neutral base in pharmaceuticals and cosmetics. Today, few people in their daily life do not wear it, apply it to their skin, swallow it in pills or read printed words from it. Out of Cookworthy's initial diggings at Carloggas, the pits swelled and deepened. They multiplied in number; in 1858, there were more than one hundred and fifty dotting this small area. Then, like the ancient fiefdoms and principalities of Europe, they began to back into each other, to merge. By 1966, they had become just twenty-four super-pits – the great white spaces on the Ordnance Survey.

Half-hidden among Cookworthy's letters in the Record Office, I found a small package. Untying its ribbon, I unwrapped it on the desk. Inside were a number of calico bags and in those were smaller packages wrapped in tissue and inside those were shards of Cookworthy's porcelain. I tipped them into my hand. They glistened in the overhead light, and when I sifted them with my fingers they made such a pleasant clink that I did it again; at the opposite desk, a researcher, hunched over a tithe map, raised his head. I bent my ear down and sifted more softly. It sounded a little like the jangle of coins in a deep trouser pocket, but it was more resonant than that. It sounded musical.

Walking down from the pits that evening, my brow felt heavy from the glare, as if I'd spent all day out at sea. Beside the old Blackpool works, I spotted a familiar van coming down the hill. It was a man who'd made a couple of door-frames for us and I leaned in through the passenger window to chat, to tell him what a fine job he'd done. His workshop was nearby and I'd once visited him there, but now, despite all my explanations, I could tell that he was having difficulty understanding what on earth I was doing up there.

I stayed in St Stephen that night and in the morning left early to look for Cookworthy's original works at Carloggas. It was a crisp March day. Haw buds dusted the hedgerows with their innocent green. A woodpecker was working at a tree somewhere up the valley, a hammer-drill amidst the lyrical chorus of thrushes and finches. The path crossed the fields after Trevear and reached the tiny hamlet of Carloggas.

The larger clay-works and spoil-heaps were all at least a mile away. Here were hedge-rimmed fields dotted with sheep and, from the yard of Carloggas Farm, a mechanical *clug*... *clug*... *clug*. An elderly man in a boiler suit stood with one boot up on the rail of a gate, watching a belt drive a churn in his milking parlour.

'Cookworthy?' He shook his head. 'Don't know nothing about that.'

'William Cookworthy? The Quaker – discovered china clay.'

He pushed up his cap and scratched his forehead. 'Well, they do say he was here. In conversation, like. That's what they say. But 'twasn't this house.'

Two hundred metres on up the road was another farm – Carloggas Moor. In an open-sided barn a man was galvanising the hubs of an old tractor. He said that a woman had come here once and told him it was where Cookworthy stayed. 'She was from America, so probably she was right.'

We looked at his tractor.

'I was all for chucking it, but she wants it kept.' He flicked his chin towards the house, where his wife was watching from the kitchen window. 'Thing is our son loved to play on it and she'd be watching him, looking out at him from up there. Now he's died – he was handicapped – she says to keep the tractor, she wants it kept, for the memory.'

He held the jar of zinc paint in one hand and with the brush pointed out towards Meledor clay works. 'We used to stand here, him and me, and you could see the sea then, out Newquay way – but now there's only the heap there.'

We stood together for a moment, in silence.

'I miss him too,' he said. 'He was a good boy.'

I went on my way, up a stony lane towards the dome-shaped hill of St Stephen Beacon. In the field below were the ruins of a smallish building with high windows and no roof. The ground beside it was broken, exposing the white beneath – I scuffed it with my boot and it crumbled into chalky dust, a desiccated version of what filled the creek below our house. The stunted thorn bushes around the ruin were hung with twists of sheep's wool. A little further up, I passed the mossy stubs of some ruined cottages, before coming out on the open slopes of the hill.

Looking south and west, down over the Fal Valley, I could see grey spurs dimming in the haze. From one side came a distant clunking from the Goonvean pit; above, the high notes of lark-song. For miles all around the land was covered not only with the gougings and moundings of two hundred years of extraction but with all the secondary activity that had sprung up – the dries and kilns, the water-wheels and leats, the rotative engines, the despatch yards and ancillary works and brick factories. Then came towns of terraced housing raised on bare fields and, in an abandoned pit on the eastern edge of the area, the biomes of the Eden Project, which itself required the erection of office buildings, car parks and new roads. Yet here, where it all began, nothing. It's as if the ghost of William Cookworthy – the quietist Quaker who called the earth our 'heavenly inheritance' – has made sure that Carloggas should look exactly as it did when he made his first diggings.

Cookworthy's perfection of porcelain in the 1760s – the experiments, the gathering of materials – appears now as the pioneering work of a scientist. But he was governed by a more traditional mindset. Like Isaac Newton, he believed in the wisdom of the past, the *Prisca Sapientia*, the notion that in antiquity there existed a clear understanding of the world which had since vanished. Scientific inquiry was not so much a pursuit that would in time reveal nature's laws, as a process of *re*-discovery. Like Newton, Cookworthy also retained a semi-secret fascination with alchemy; most of the marked pieces of his Plymouth porcelain carry the alchemical symbol for tin.

William Cookworthy was also a great advocate of 'the rods', of divining. He learnt the practice from a man called Captain Ribeira, who had deserted the Spanish navy and ended up as commandant of the garrison at Plymouth. Ribeira was quite happy to show anyone who was interested how to use his dowsing wands. What he refused to do was to tell them how to distinguish particular ores. But Cookworthy worked it out on his own. He wrote an essay on divining in

*Mineralogia Cornubiensis*, explaining his intent: 'simply to set down what I have observed, in hopes that the Instrument may come into use and be of service to mankind'. But the essay is far from simple. Cookworthy had taken Ribeira's knowledge, added what he had read in Georgius Agricola's *De Re Metallica*, and spent a great deal of time in the field. He outlined how a multi-layered subterranean map could be drawn up, distinguishing the ores, charting the rich lodes and the watery voids with instructions like: 'AA is the Vein; GGG, the Flookan, B the place of Interfering…' Cookworthy's essay is one of the most comprehensive texts ever written on dowsing.

When Samuel Johnson came to Devon to stay with Joshua Reynolds in 1761, Cookworthy offered him a demonstration. The two Londoners were sceptical but they agreed to hide a druggists' mortar in a couple of feet of soil and Johnson watched closely as Cookworthy paced around the garden looking for it. When Cookworthy felt the twitching, they dug. Nothing. The mortar was retrieved from elsewhere and Johnson scoffed at this credulous Quaker in breeches and tricorn. But Cookworthy inspected the mortar and said, 'It is bell metal [an alloy], the mixture has destroyed the native metal.'

Much of Cookworthy's scientific enquiry was more orthodox. He studied astronomy (it was in his library that Daniel Gumb, the wild cave-dweller of Bodmin Moor, learnt much of what he knew of the stars). He devised an elaborate method of desalination to prevent the Foulwater Fever – a common shipboard ailment. He was among the first to identify the cause of scurvy and recommended stowing sauerkraut, rich in Vitamin C, for long sea-voyages. When the engineer John Smeaton came to Plymouth to rebuild the Eddystone lighthouse, he lodged with Cookworthy, who helped to develop the particular lime that made his structure so robust.

As a Quaker, Cookworthy was driven by the doctrine of 'inner light'. His work as a minister was to help others to uncover this light amidst the layers of earthly desires and doubts, just as his dowsing revealed living metals beneath the base soil, and his work as a chemist rid the dross from pure substance. What is remarkable about William Cookworthy is that although he explored both matter and spirit so deeply, he never saw the two as separate. He resisted the presiding Cartesian dualism of the age, seeing the world not as a machine but as a book of wonders to read and interpret.

When he first came across the esoteric writings of Emmanuel Swedenborg, Cookworthy is said to have hurled the book to the ground. But in time, he became so convinced of their truth that he translated and published *The Doctrine of Life*, the first of Swedenborg's works to appear in English. Like Cookworthy, Swedenborg was a chemist and mineralogist and he gave expression to the idea that the elements are not entirely inanimate. Of the second chapter of Genesis, when God forms man from the dust of the earth and breathes life into his nostrils, Swedenborg wrote: 'To "form man, dust from the ground" is to form his external man . . . To "breathe into his nostrils the breath of lives" is to give him the life of faith and love'.

George Fox, the founding father of the Quakers, taught from the same Biblical passage that the breath of life was part of a divine substance which formed the soul of each one of us. For Cookworthy, both alchemy and the quest for true porcelain had their analogy in this text – the transmutation of the soil beneath our feet.

The earth lives. The land is filled with spirit. Having just discovered both china clay and china stone here at St Stephen, Cookworthy wrote to a friend about the task ahead: 'I have the flesh, the bones and the sinews but to put flesh on bone to make a whole body, that is what perplexes me.'

# 9 | FAL

Obscure, possibly pre-Celtic, although Ptolemy mentions the Cenion river in the first century AD, which appears to refer to the Fal.

I WAS LOOKING FOR THE SOURCE, climbing fences, fighting through thickets, squelching across bogs, wrestling with the map. Must be here somewhere . . .

At little more than twenty miles long, the journey along the Fal, which I was now boldly undertaking, is hardly one of the world's great river journeys. But the size of the Fal is not the point. All rivers are stories – connecting places, carrying history – and on those terms, the Fal scores pretty well. In antiquity its waters led ships to the heart of one of Europe's most mineral-rich areas; a tin ingot from the fifth century BC has been found at its mouth. The river is mentioned by Strabo and Pliny and Ptolemy. For centuries, its creeks offered shelter to early Christian ascetics, medieval royalty, and the mercantile and naval fleets that shaped the modern world.

The Fal's length has fluctuated wildly, chopped down and extended with each passing age. Twenty thousand years ago, with ice sheets covering the north of Europe and man pioneering at their fringes, the sea was over one hundred metres below the current level, and the river pushed far out into the English Channel. By the time of the Romans, the Fal Valley was flooded and the tide pushed a lot deeper inland than now. For two thousand years, the acceleration of man's activities – tin-streaming, farming, digging for china clay – led to it silting up, pushing the tide back out to sea, adding a third to the river's length.

So, I was looking for its beginning. I'd always thought that it was on Goss Moor, but on the map a stream ran into the moor from the east. Following it up towards the clay-pits takes you to a place called Pentyvale or *fynten Vale* – where *fynten* is 'spring' and 'Vale' is a sixteenth-century form of 'Fal'. Pretty straightforward, you'd think – but, here on the ground, it is clear that clay extraction has affected its rising just as it has the rest of its course. The longest stream comes now not from Pentyvale but from the man-made slopes of Littlejohn clay works.

Clambering down through a brake of blackthorn and bramble, I reached the edge of a small pool. High above it rose a thirty-metre-high barrage of landscaped spoil, snaking on out of sight for miles – and from its base, between mossy boulders, a glassy arc of water. Here was where it began – at a new spring, in a new slope. I took off my boots, stepped into the pool and cupped my hands, sipping at the water and chuckling at my seriousness.

Below the spring stood a bungalow, and a Cornish flag on a bamboo cane. In the adjacent field was the abandoned hull of a ski boat. Within an hour, I had reached Goss Moor and a sudden expanse of marshy ground, covered in low alder and willow, where the river wanders about like a stranger in fog. Here minerals leach bright rust into the water. In several places I spotted otter spraint and claw-marks in the muddy bank. On the far side of the moor is one of the Fal's few stretches of footpath. Joining it, I encountered a long line of teenagers on some sort of organised hike. I passed several groups, each with its own dynamic of jostling and ribbing and flirting.

'Where are you going?' I asked.

'We don't know!'

Two girls had stopped to peer into a culvert. One sucked at her pack's camel-straw. 'Ye-uck,' she was pointing at a ferrous slick of orange.

'Looks like puke,' said the other.

At Trerice Bridge, I paused to let them all file past me, up the road towards St Dennis. I looked at the map. For three or four miles now the river was squeezed between clay pits and old works – no sign of any path. I climbed over the wall, and dropped down into undergrowth. Dried sticks snapped beneath my feet. The canopy echoed with bird-song. The river had gathered the waters of numerous streams and seepings and was now several metres across, pushing through a V-shaped valley. I walked slowly, keeping my eyes ahead; it all looked thorny and abandoned and I imagined I'd soon be forced back, up into St Dennis, to eat a pasty outside the bakery with all those reluctant hikers.

For a mile or so, though, I managed to stick close to the bank, pushing aside head-high growth, crouching and crawling at times beneath tumbled boughs, picking brambles from my shoulder, wading side-channels, treading narrow and crumbling banks until – whoosh! – I was up to my knees in the river.

I climbed back up, over an old concrete wall. I took off my boots and wrung out my socks and sat in the sun to dry out. Behind me stood a group of unroofed buildings. Ivy curled out of paneless windows; a rhododendron pushed puce-pink from a deserted workshop. I watched a peacock butterfly tangle briefly with an orange-tip before realising its mistake and flying off at speed.

Above my head ran a large rusting pipeline, mounted on a series of high concrete piles. I followed it along the bank, through budding scrub, trying to work out its function. Every twenty metres or so was a wheel-tap – twist it and the solution would have gushed into the allotted drying pan. I stood in one of these pans, the low concrete wall a tangle of ivy, and imagined it filling with the porridgy liquid. 'Industrial archaeology' always seems to me the wrong term for studying such sites. The evidence is too complete; it lacks all the gaps and enigmas of prehistory.

Water was part of every stage in the production of china clay – it still is. The clay is washed from the rock with high-pressure hoses, then pumped up from the pit to run down through a labyrinth of pipes, launders and drags. It takes weeks to rid the water of grit and sand and mica, and leave just the fine white china clay, dried into press-cakes and ready for despatch.

Each of the dozen or so pans was about the size of a tennis court and they were filled now with the pioneer growth of buddleia and elder and sycamore whips. They were well on their way to being woodland again. The concrete borders were crumbling like old biscuit. No one had been here for a long time: at every step my chest snapped off dry twigs, and cobwebs brushed my face.

At the river's edge, the wall fell away and I leaned over and saw the water some six metres below. Another series of pipes pushed out from each pan, probing space. Their mouths were dry now, and crusted with rust. But for years, they had spewed a solution of quartz granules and mica sand into the river below, and I remember as a child seeing it far downstream running like milk beneath the bridge at Tregony.

'A landscape of purgation,' wrote the Cornish poet Jack Clemo – all that water, all that rinsing. Clemo's image isn't just a literary trope; he really did see his native clay country as a place of soul-washing, an earthly purgatory. Given his story, it is not hard to understand why.

Coupled with his own physical handicaps, the area's man-made topography produced in Jack Clemo something of a visionary. In a letter written late in life, in 1987, he explained: 'The clay landscape dominated me from childhood, and from my teens to my forties my reactions to it were instinctive, complex and apocalyptic.' His friend Charles Causley thought him a genius, 'one of the greatest' writers Cornwall has produced. Others saw him as the last of the self-taught working-class poets, in the tradition of John Clare and D. H. Lawrence. His response to the scarred earth around him was pitiless and austere. His voice is the conscience of the post-industrial age, crying from the white wilderness of Cornwall's clay country. As Cookworthy did, Clemo saw the land as a manifestation of the divine. But Cookworthy saw God's providence in its hidden properties, whereas Clemo – with the scars of production all around him – saw only wrath and vengeance.

Jack Clemo was born in 1916, in a cottage not half a mile from Cookworthy's first china-stone mill at Carloggas. He lived there for the next sixty years. The slope of Bloomdale spoil-heap rose just metres from the back windows. Jack had no brothers or sisters. His father had been killed in the war before he was born. Most of his childhood was spent alone with his penniless mother. The difficulties of such an upbringing played their part in Clemo's life – but less so than the hammer-blows of his physical afflictions. For long spells in his youth he was blind; at nineteen he became completely deaf and remained so for the rest of his life; in his late thirties the blindness returned and stayed for good.

White pyramids, chasms and pits, stent-mounds, the whir and hum of machinery, the boom of blasting – these were the shapes and sounds he remembered from the years before the blindness, before the deafness. In his autobiography he recalls what he witnessed early on: 'The soil was thrown into tanks and kilns and it brought to the human spirit more poignantly than anything in the peaceful countryside the sense of insecurity, the sudden pounce of the destroyer.'

Throughout his childhood, when he could see, Clemo had watched the march of the diggers, the widening and deepening of the pits and the growth of the spoil-heaps beside them. Open ground that he walked one month would be gone the next. Flower-thick hedges were buried or grubbed up, along with centuries-old fields and paths. It was, he said, the very opposite of the repeated patterns of the natural world, the perennial cycle of decay and replenishment: 'There were no rhythms about it, no recurrences; only a pitiless finality in every change.'

Yet Clemo's view was more nuanced than the dualism of industry (bad) versus nature (good). For him, a perverse truth lay in the 'fantastic swellings and angles of the clay landscape'. All around him he saw not the wanton discardings of industry, but a fitting topography for the Fall, the contours of man's corruptions. It was of little regret to him that the land had been ripped open and scarred. He had no time for the flowery gaze of the Romantics and dismissed the 'nature worship' of the likes of Shelley and Jefferies and Thoreau. Clemo's God never intended anything pleasing in Creation: '*His* hand did not fashion / The vistas these poets admire, / For he is too busied in glutting / The worm and the fire.' Those who glimpsed the eternal in the earth's beauty displayed for Clemo 'a slackness of fibre'.

He was only four years old when he first began to complain of a pain in his eyes. It became hard to tolerate any kind of light and his mother grew accustomed to finding her son cowering in the darker corners of the cottage. Within weeks, his eyes were so swollen that he was unable to open them. So began a year of darkness and of fruitless trips to doctors. 'It became,' he wrote, 'a familiar and pathetic sight to the people of St Dennis, the solitary widow, dark, tight-lipped . . . wheeling down the hill from Rostowrack the sickly boy whose face was swathed in bandages.'

That first bout of blindness lifted of its own accord, but Jack was for ever changed by it. Once boyish and robust, he was now frail, thin and bookish, an isolated child in his high, isolated cottage beneath the clay-dump. As he turned thirteen, the affliction came back. This time he developed such a horror of the linen bandages that, rather than wear them, he slunk off to the top of the stairs. There, in darkness – 'like a captive bird' – he squatted on a shelf.

He rarely ate or spoke. He spent the time in meditation, ridding his mind of the rote of his schooling. In its place grew a hard-edged moral geography constructed around the places he remembered. When his sight returned, he discovered the familiar world underpinned by a nascent faith. Over the coming years his beliefs were moulded and broken and remoulded as he tried to keep pace with the full fury of his sensitivities. By his twenties his beliefs had grown more solid as he developed his own brand of Calvinism. In later life he softened a little, but the rewards of his faith were never earthly comforts: 'I realised that a natural mystic is no closer to Heaven than a natural materialist.'

Jack Clemo would have recoiled from being considered a poet of place. The ore of his work came from within, from the lightless and soundless shafts of his own spirit. Yet it was the land around him that became the grist for his writing. He returned to it again and again. Clay was the raw material for his poetry, and proved endlessly malleable. In his collections *The Clay Verge* and *The Map of Clay*, it recurs again and again. He writes of the 'clay-world oasis', 'the clay altar', 'these weird aisles of ghostly stone'. The features of the clay country become abstracted, and abstract notions become its features. Hensbarrow is a 'clay-smirch'. Hope is a 'cone of dreams, / White cone, treasured where the mine is open'. Faith is a hard 'creed-cone' and religious striving the clay-man's quest: 'Men burrowed here but found no clay.' He equates himself with the land: 'My character, like the scarred plateau that bore me, had its geometry askew.' He is a 'clay phoenix'. Clay is sex – 'a dry cool breeze exposed her clay', 'lest my male clay be hurled / To flame . . .' while the clay-works were used as a stage for its absence: 'I track dry kiln-beds, miss the lure, / and slink unpurged through stale dust-laden air.' And clay is vision. From the 'venus white clay glare' emerged blurry figures and hallucinations. In 'Christ in the Clay-tip' he recounts: 'I peer upon

His footsteps in the quarried mud; / I see His blood in rusty stains on pit-props, waggon frames / bristling with nails…'

Clemo transformed the land around him into a mythical place, a Dantesque arena of progressive intensity. In *The Two Beds* he writes of D. H. Lawrence, whose passionate questing failed to deliver the gleaming faith that Clemo himself had discovered. Born in the coal regions of the north, Lawrence remained 'a child of the black pit'. Clemo, though admiring of Lawrence's work, pitied him his Godless gropings: 'You never saw / The clay as I have seen it, high / On the bare hills, the little breasts / So white in the sun.'

For the rest of that morning I pushed on beside the Fal, struggling through the bankside tangle of new growth. The river disappeared between clay workings, squeezed and part-dammed by earth-shiftings, an intruder in its own valley. I too felt like a fugitive, ignoring the STAY OUTS and NO ENTRYS, hurrying across tracks and open areas, hiding from the yellow works Land Rovers. They'd been blasting at Wheal Remfry and through bare branches I watched the trucks toiling up the pale roads, their wheels and high sides pale with clay-dust, beneath the pale cliffs.

Before Treviscoe Woods stretched a deserted drying shed – the last stage of the process that took the clay from rock face to truck, where the residual water was heated away to leave the press-cakes. The shed was a hundred metres long, open on one side beneath the high concrete shell of the dries. Walking through it, my footsteps echoed off the concrete. I could see the waste ground outside bright in the sun, then dappled beneath the riverside trees. The shed was so full of reflected light I had to squint.

At the far end were two great furnaces. The rusting doors hung open and inside them the fire-bricks were black and flaky with soot.

A broken stairway led up around the furnaces, and I toed each of the steps before trusting it. Coming out on top, I looked back along the length of the dries. Honeysuckle and bramble had snaked in among the frames where the clay was heated. From a patch of mulch in one of the cavities, a lone birch had taken root. Its slender stalk rose straight before suddenly shooting off towards a hole in the roof-sheets. I thought: in years to come, when the roof has gone and the birch is fully grown, you will be able to calculate the start of the roof's collapse from that kink.

I spent a long time in the shed, amazed by its scale, its redundance. It was if some marauding army had marched through, driven everything before it and then hurried on. In the far corner three bus seats ringed a pile of ash and half-burnt logs. Beside it was a sun-yellowed copy of *The Daily Mirror*. It was dated 22 May 1999.

Back outside, I carried on downstream. The river pressed on into woods. The metallic clanking of the clay-trucks grew muffled and distant. Clear water rolled over rocks; a jay-screech jabbed somewhere high in the chamber of trees. Treading the leaf mould, watching the patterns of sun and shadow on the greening ground, it was almost possible to forget the industrial scars and wallow in the thought of spring. But I had Jack Clemo with me, and his words burned in my pocket: 'I am beyond your seasons: food / For these is in your blood but not in me. / I lapse from Nature towards a birth of Heaven's fertility / That blasphemes Spring upon your earth.'

Clemo was fiercely aware of his exile, his exclusion from what gave others solace. Yet for all his aversion to natural beauty, he described many of his most intense moments in relation to the places where they occurred – often on top of Bloomdale clay-tip. It was there that he first 'dreamed of strange gods', where later, in his teens, as his sight began to come back, he saw butterflies blurring with the gorse and the flashing mica in the spoil. In the summer of 1938, devastated by the end of a passionate attachment to Barbara Rowse, he climbed Bloomdale to kneel on its peak. This time his eyes were blurred by tears; he grappled with the twin forces that surrounded him – 'the tide of prayer . . . and the black tide of lust'. He returned exhausted and placed under his pillow a lock of Barbara's hair.

Following their break-up, he took to roaming more widely, praying in deserted clay-work huts, with their cobwebs and splintered boards. In his despair, he developed a horror of all that was whole. During the war years, he found another spoil-peak to climb at dusk and there kept a 'lonely vigil until darkness fell on the blacked-out village and Carne Hill became a blurred snout'. But now there was something new – not the sexual anguish and doubts of old, but a peace that he saw spread out across the landscape before him: 'It was a pure and unchallenged spiritual tide that not only flowed through my soul but thrilled and pulsed as a physical sensation'. He was sometimes filled with such a deep and overwhelming joy that 'the tears would stream down my cheeks as I moved through the rain-soaked bracken or up the wet sandy slopes'. Without these exultant moments, his poetry and his faith would not have survived the dark years ahead.

Jack Clemo developed a strong belief in the notion of 'election'. His difficulties had been placed before him by God; his given task was to transcend them through 'the mystical-erotic quest' for a partner, through writing, and through religious belief. Each of these ideals he pursued with the same fierce courage, and with each he was hampered by the loss of his senses. From his teens onwards he was hit hard by a number of acutely felt and ultimately unsuccessful emotional attachments. No less easy were the thirty-seven rejections he received before his first novel was published (he had written eight novels by the age of twenty). But most arduous of all was the route to spiritual fulfilment, a wide and erratic orbit around his received Methodist faith and the four walls of Trethosa Chapel.

*

Just half a mile from his cottage, the chapel stands alone on a knoll above the Fal Valley. I climbed there from the river, along old pit tracks and scrubby new woods. Its grey walls and grey slate roof rose against the mountainous white of the dumps behind it.

The key was kept by a Mrs Luke; I had called her earlier. When she came, the lock stuck. 'Sometimes, it just needs a little shove. You try.' I pushed once... twice, and the door sprang open on to a large room smelling of stale air and damp. Black-and-white pictures, trestle tables and hard brown chairs. The ceiling was low, as if the floor above was pressing it down. Up there was the chapel, and I followed Mrs Luke through a kitchen and up the flight of wooden stairs, which rose into a screened-off side aisle.

'These screens can come down,' she explained, 'if we need more seats. But it doesn't happen much nowadays. Last time was a funeral – just the other day. That was the first time since 2000.' She sighed. 'Young people now just seem so busy.'

I followed her into the main room. A large organ dominated the chancel. 'That,' she said with a slight straightening of her back, 'is one of the finest organs in Cornwall.'

'Do you remember Jack Clemo?'

She nodded. 'He was a well-loved man.'

'But he didn't always see eye-to-eye with the chapel.'

'Well, that's true.' She hesitated, as if for her that wasn't really the point. 'He was not straightforward. I have to admit that a lot of his work goes rather over my head.'

After his cottage, the chapel was probably the most significant single place in Jack Clemo's life. In 'Beyond Trethosa Chapel', he writes of the efforts to square Methodism with his own experience, to fit the plain-spoken wisdom of the chapel in with the flares of his spoil-heap ecstasies: 'It flashed in Cornwall, at Meledor / My rebel vision, kindling the scarp'. For many years, he was estranged from Trethosa and its congregation. At chapel times he would slope off to his haunts among the spoil-mounds: 'I preferred a rough claywork cuddy of rusty zinc to a chapel.' Sometimes on his wanderings he would spot in the distance its grey walls and grey roof: 'The Bethel stood in full view, a sharp / Alien scab across the dale, / On the fork of the hill: its lure was stale.'

The poem predicts that one day he would find his way back to the fold, with the help of a woman: 'Mediate, then, beloved; let tension cease, / Dune-grit and pews be reconciled … / Bless with your dreams my broken clay / As you take the broken bread.'

In his thirties, he took to going to chapel again. He was still unmarried, still living with his mother beneath Bloomdale clay-dump. For years, the two of them walked together to Trethosa, usually twice a week. They sat side-by-side in the same pew. As his sight faded again and he could no longer even lip-read, Jack's mother learnt to spell out on the palm of his hand the words of the preacher. He said he was able to 'hear' the notes of the organ through the vibrations in his feet.

Mrs Luke led me down the aisle to a bench three or four rows from the back. 'Here, this one – this was his pew. His and his mother's.'

Clemo did find love. At the age of fifty-two, he married Ruth Peatty, an attractive woman from a devout Plymouth Brethren background. She had written to him out of the blue when she heard about his work. Clemo responded by sending her a copy of his *The Invading Gospel*; the relationship blossomed from there. The wedding took place here at Trethosa chapel and Mrs Luke remembered it. 'A lovely occasion. The chapel was full to bursting. We all stood outside after and had our picture taken. Very wet it was!'

Clemo lived on in the clay country with Ruth for a few more

years, before they moved to Weymouth. He died there in 1994, at the age of seventy-eight. 'From middle-age onward,' he wrote, 'I lived in the summer sunshine of fulfilled hopes and answered prayers, amid the ripening comforts of marriage, friendship, church fellowship and prosperity.'

Below the chapel at Trethosa was a small room behind a door of frosted glass. A wooden plaque above the door read: *Jack Clemo Memorial Room*. Inside two fluorescent tubes flickered on and off, and on again, before shedding their hard light over a row of display panels stuck with pictures and poems, newspaper cuttings and obituaries.

'It's not much,' said Mrs Luke. 'But there's nowhere else really, nowhere to remember Jack like this.'

He was there in a number of photographs, distinctive in beret and dark glasses, flanked by his mother, or later by Ruth – and in one picture by both. And there was the group standing outside the chapel on his wedding day, the slate flags shiny with rain.

His desk stood in one corner, a small bureau in plain hardwood. A song-book lay on it, along with a large family bible. Beside it was a pair of gloves and a coiled-up belt. 'Those are all Jack's,' said Mrs Luke. Beside them was a polystyrene hatter's mould. On it, set at the same angle as in the photographs, was his beret.

The room was dominated by a model – a 1:12 reconstruction of the Clemos' cottage. It was beautifully done. Fourteen hundred hand-cut slates covered the roof, and the walls were painted in delicate render. I bent to peer into its scaled-down interior. My own eye was briefly reflected in a stamp-sized pane. Behind it was a miniaturised version of the space where Jack spent the greater part of his life, where he wrestled with his writing, with poverty (he had had to give up his occasional journalism because he couldn't afford the postage), with the burden of his 'election' and the painful press of the light. But, like him, I saw only darkness.

Over the course of his lifetime the places he knew disappeared one by one, covered by the white creep: Rostowrack Downs, which he crossed in a push chair with his eyes bandaged – *gone*; Meledor, where he'd had his visions – *gone*; each of the houses where his parents grew up – *gone*. When he died in 1994, the cottage still stood. But just a few years later the Goonvean works expanded and they put a demolition order on it. Clemo's supporters in Cornwall at once applied to preserve it. Clemo's stature had grown – from a combination of his work, his life and the fading version of the Cornwall he represented. They offered to move the cottage stone by stone to the China Clay Museum at Wheal Martyn. But before the matter could be resolved, the cottage was pulled down.

'They did the dirty,' Mrs Luke said, looking down at the model. 'They did the dirty on us.'

It was odd – the scaled-down model suddenly looked scaled-up in that small room, filling it with a fulsome presence that the cottage itself had lacked as the clay-dump grew above it. Preserved here, it made me yearn for a man I'd never known, whose work now coloured the spreading white space of the area. But I couldn't help feeling that Clemo himself might have smiled at the cottage's demolition – the young Clemo crouched in an abandoned clay-hut, raging against his physical impediments, against nature-lovers and the sentimental, against chapel-goers, the Clemo who saw God's will in the spoil-smothered pasture around him and derived a perverse pleasure from devastation, who wrote:

> Praise God, the earth is maimed
> And there will be no daisies in that field
> Next spring; it will not yield
> A single bloom or grass blade.

# 10 | TREGONY

*Tre-*: Cornish, 'farm'; *-gony*: possibly from a personal name, 'Rigni'. The stress on the first syllable, *tre*-gony, rather than on the more usual second syllable, tre-*go*-ny, suggests that the 'o' is added or epenthetic. There is no obvious root, though, for the name 'Rigni'.

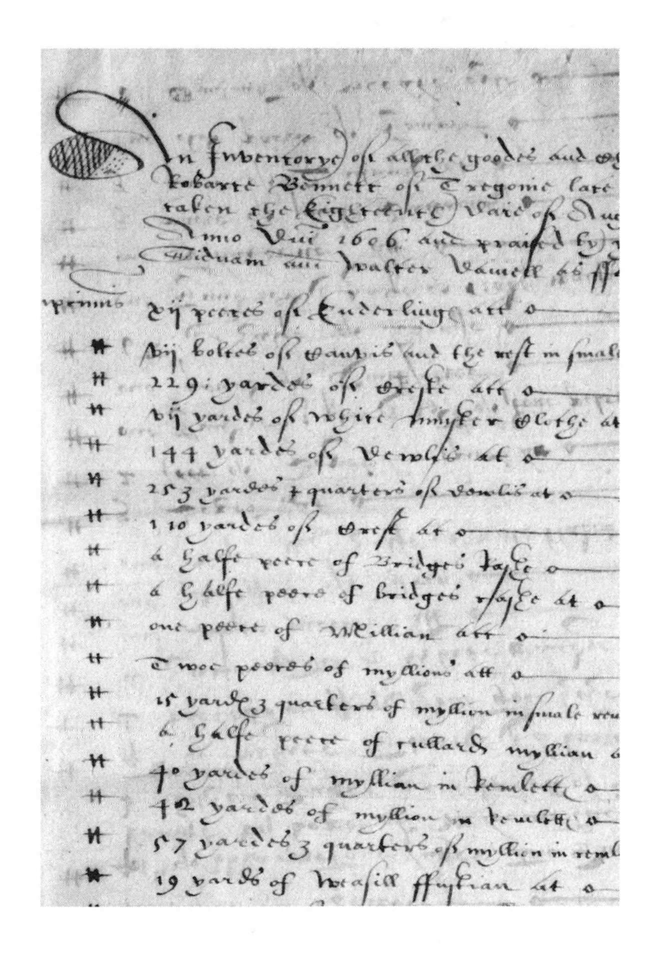

Below trethosa, the fal wound past the last of the mica dams and at St Stephen Bridge left the clay country for good. I crossed a meadow and stood in the shade of some bankside willows. Around my feet was a burst of spring growth – white stitchwort, pink campion, yellow celandine, bluebell, a dozen different greens. The ferns stood stiff amongst them, their crozier coils set to unwind. The river here was still no more than four or five metres wide. Against the far bank, where the flow was faster, a long trough of water dipped and rose to a back-wave, constantly breaking, filling the canopy above with its miniature surf-sound. Downstream, in the evening light, I could see the water shining beneath a swirl of midges.

The transition had been so sudden that I had the sensation of being spat out – out of the white of the gorge above, the old clay workings, the empty halls of the kiln-dries, away from the background racket of the still-working pit at Meledor. The river's story so far had been of the centuries of extraction. Now I watched it gliding on down the valley with nothing busier beside it than the slow revolving of bovine jaws. Only Brunel's viaduct, stilt-legged in the late sun, was a reminder that the nineteenth century had taken place at all.

Some way beyond the viaduct, I thwacked down a patch of riverbank nettles with a stick and pitched my one-man tent. The night was a pleasant gabble of water from below, the rattle of a couple of distant trains on the viaduct and something long-legged splashing across the river.

In the morning, I pulled the flaps open and watched the first of the sun yellowing the oak-tops. Pearls of dew lay on the old grass underfoot and ran off the tent as I dropped it, in worm-like globs.

Ahead the valley was a patchwork of sloping fields and flat meadows – but all on the other side of the river. The only place to cross was an abandoned weir, its iron girders rusted to tapering struts. The timber frame was rotten and a five-bar gate placed flat over that was also rotten. I took it slowly, inch by inch, creak by creak. It was something of a relief to jump down on solid ground, to step out through the mud ridges of a gateway and into the valley.

It was one of those spring mornings when you wonder what could possibly be so urgent as to keep you from doing exactly this, first thing on every such day. There was a chill in the shadows, the leaves were small, the larks sang in a cloudless sky. I stopped to inspect a knee-high sycamore and its tiny red-brown leaves. The chlorophyll was not yet in their flesh, but the outer leaves, the earlier ones, were as big as a man's hand now and floppy, as if they could not keep up with their own eagerness.

The morning's joy was undimmed even by the thought of Jack Clemo, out walking his dogma, scoffing at all that 'springtime loveliness idolaters adore'.

At Grampound is the main road, and a bridge; the name of the town, from '*grand pont*', speaks of the river's ancient overseas connections. Below it, the traffic noise dimmed in my ears and the Fal settled back into the important task of flowing gently. The valley sides deepened, the water widened, the woods thickened. Apart from the village at Tregony a couple of miles down-river, the main channel now runs on largely unseen, through unpeopled acres of tangled carr and salt-marsh, on past the tide-head and the mud flats below, between pathless oak woods, until, some twelve miles from here, dog-legging past Turnaware Point, it joins the wide estuary of the Carrick Roads, and the open sea.

Yet for two thousand years, and probably longer, the banks of the Upper Fal were a-bustle with activity. Punts and barges criss-crossed the river, netters and mussel gatherers worked its foreshore, fullers dipped fleeces in the creeks, lime-burners stoked riverside kilns. Larger trading ships were rowed up to Tregony to land wares from the continent and the Mediterranean, and fill their holds with wool or streamed ore. Ptolemy may have been referring to Tregony when he included 'Voluba' in a list of ports. The village had, at various times, a sizeable motte and bailey castle, a large manor, a prison for 'lunatics', and a great number of alehouses. Its priory cell of Augustinian monks was endowed by the diocese of Bayeux. The church of St James's, near the quays, suggests in its dedication a pilgrimage point for the route to Santiago de Compostela. Before the silt built up at its doors and forced the church's abandonment, its strong-boxes had a greater cache of silver than any in the entire deanery. In his *Historical Survey of the County of Cornwall*, W. Penaluna noted that in the Middle Ages, Tregony was the principal port on the Fal, which itself 'excels all the harbours of the isle except Milford Haven'.

A few miles on, at Ruan Lanihorne, stood an even larger castle; in the whole of Cornwall, only Launceston's was larger. A third castle was said once to have stood at Trelonk, another perhaps at Ardevora. In the names of riverside fields are whispers of vanished houses, side-creeks are called after water-wheels and tucking-mills. Abandoned quays and tree-filled quarries dot the shore. At Trelonk, the lone tower of a factory stands beside the water, where the clay spoil was recycled from the riverbed to bake into building bricks. Opposite the old factory is the creek of Lamorran, where thirteen centuries earlier St Moren established a monastic settlement, later one of the area's wealthiest manors. At Ardevora was another large manor and the chapel. Downstream from Tregony was the village of Sheepstall, with its own quays and mills, a lazar house and a chapel.

Not one of these places now exists – not Sheepstall, not the castles or the kilns or the quays. Even the ruins have disappeared. Only in a few written records does any sense survive of all the commodities, all those venturers and seamen, all that pilgrim devotion, all that *life*. One such ghost is Robert Bennett – a Tregony merchant who died in the early seventeenth century. His will survives in the Record Office and with it an inventory of thirteen pages of neat, tea-coloured court-hand detailing a range of four hundred 'shop' goods. Of all the merchants' wills from Cornwall at that time, Bennett's of Tregony is by far the longest, and widest-ranging. He held in store sixty different types of cloth, including canvas, crest, bridge rash, lace, coloured osnaburg, black jeans (Genoese), green buckram (originally from Bokhara), cambric (from Flanders), silks from Bologna, Naples and Spain, fine white calico (originally from Calicut), cushion canvas, crimson moccado, tuft fustian, tuft sackcloth, tuft leven (from the Levant), holland, fine holland, poldavies (Breton sailcloth), and tabby (a silk taffeta which takes its name from a district of Baghdad). He kept a stock of stavesacre, a substance derived from *delphinium staphisagria* (from southern Europe and Asia Minor), used both as an emetic and as a rat poison. Among the foodstuffs he could offer were cloves, aniseed, ginger, prunes, treacle, figs, raisins, currants, vinegar, isinglass, fenugreek, mace, almonds and rice. As a condiment for the beer sold in Tregony's numerous alehouses, he carried the spice derived from capsules of *amomum meleguetta* from West Africa and known to those lucky enough to procure it as 'grains of paradise'.

Soon after midday, I reached Tregony, now little more than a large village. In its Londis store, I found a bottle of water and surveyed the shelves – Crunchy Nut Corn Flakes, Chilean Merlot, bananas . . . Bennett certainly had more cloth on offer than Londis, but no bottled water; otherwise his range of four hundred items, from many countries, is comparable. The difference is that Bennett's wares were all likely to have been landed within a few miles of here, and would have lacked the elaborate packaging.

I sat on a bench outside and drank the water. Not much of the traffic stops now in Tregony – most just drives through on its way to somewhere else. Nor do many people consider why the main street is quite so wide, nor realize that the rough ground just below the town was once a gateway to the rest of the known world.

A small lane wound on beside the river, past the wooded place where Sheepstall had once been. I left the lane for a grove of high beeches, and, out of a fierce loyalty gathered from all of two days, stuck to the riverbank. I followed a shooting path and the woods thinned to steep fields terraced by animal tracks. Then the valley bottom changed, the river swung away and I entered a wide expanse of bog dotted with iris and alder.

I imagined I'd be halted here, forced back by mire or a side channel too wide to leap. It was strange country, caught in the last stage of the silting process – the transition from mud flats into soil. According to the *Nature Conservation Review*, it represents a habitat 'almost extinct in Europe'. It was to see this that Richard Mabey reports crossing the country some years ago, drawn also by the promise of one hundred and twenty flowering plants and the phenomenon of 'high tide in a springtime wood', primroses growing under sea-water. He saw many of the plants and the high tide, but the submarine primroses proved too good to be true.

Ahead I could make out the three spans of Sett Bridge where, at high water, the Fal ends its brief freshwater existence. Now, with the tide out, the water was sluicing over a weir-like apron beneath the arches, on between grassy banks, through another two miles of mud-fringed and wood-flanked valley – another hour or so of extended life. I leaned on the parapet and looked down into the vast basin of Ruan Creek, a mile across, tens of metres deep in spoil. I thought of the gouge of the Melbur clay works and wondered how the two compare in volume. A lot of earth has passed under the bridge since William Cookworthy first filled his little bag at Carloggas.

The water runs clear now. By the 1980s, the pits had been banned from discharging waste. Already the mud is beginning to shift. I could see to the right of the bridge that a large slab of bank had just collapsed. It lay half-submerged in the flow, swiftly breaking up. Where it had broken away, the fresh cut was as white as flour.

They were digging a grave in St Rumon's churchyard. Or rather, one man was in the grave, digging, while another was above, in red braces, busy leaning on his spade. He was an elderly man with a jowly face and was quite happy to interrupt his leaning for a little chat.

'It's for an up-country gentleman.' He nodded down into the pit.

From below, a shovel blade rose and a scatter of soil fell on the mound beside us.

'Exciting place, a graveyard. Least I always think so. Always something going on.' We looked around at the headstones and the empty paths and the shadowy places beneath the sycamore. He extended a finger to an age-skewed memorial beside us. 'Best stones are they slate ones – like that. Nice curly writing. Stays hundreds of years on slate – not like the limestone. Weather gets to the limestone and it's gone in no time, wiped away.' His voice trailed off. 'Then you're just gone . . .'

I asked him about John Whitaker.

'The parson? He's inside.'

The church door creaked open. I found Whitaker's daughter and his 'relict' wife robustly remembered on large slate slabs beneath the chancel steps. John himself – thirty years vicar here – was right up next to the altar: one oak leg rested on '*Joh-*'. His women had birth dates and epitaphs, but John's stone was somewhat curt:

*John Whitaker*
*BD Rector*
*Buried November 4th*
*1808*
*Aged 73 Years*

John Whitaker was also an up-country gentleman, originally from Manchester. He was a fine example of that multifarious figure, the eighteenth-century antiquarian, and played his own small part in the history of the nation's shifting attitudes to place and the past.

Whitaker had excelled at school and then at Oxford, was ordained and went to London. He knew Dr Johnson. He was friends with Edward Gibbon (who showed him for comment the manuscript of *Decline and Fall of the Roman Empire*). In 1771 he was elected Fellow of the Society of Antiquaries. He wrote two volumes of a projected five of *The History of Manchester*. In doing so, his intention was not to add to the 'private and dull annals' of provincial history but, like John Leland and William Camden, to reveal the glories of Britain by looking at the places where those glories happened. Such studies invariably use Arthurian lore, and Whitaker's *History of Manchester* was no exception. King Arthur, he discovered, had fought an 'uncommonly bloody' battle just outside Wigan.

It is not for his *History of Manchester*, though, that his story is told here, but for an entirely unknown work that arose from his years of Cornish exile. In November 1773 John Whitaker had been appointed the morning preacher at London's Berkeley Chapel, and used the pulpit to benefit a broad range of people with his wisdom.

Unfortunately, not all of his wisdom was very tactfully put. Whispers of libel began to attach to his name. He soon found himself eased out of his position and given a new parish – Ruan Lanihorne, in faraway Cornwall. After days of travel, he found himself looking out through the windows of his new rectory on a scene of unbroken stillness – wooded slopes, open fields and the mud of a clogging creek. Around him were people whose concerns extended little further than their own sheep and cattle, or the size of fish in the river. They certainly had no interest in Whitaker and his learning; the Church of England was much too patriarchal and English to accommodate their beliefs. Only seven people took communion at his first service. Whitaker looked at the empty pews and realised 'with sorrow their aversion from his teaching, their indifference to his instruction, their repugnance to his authority'.

Such sorrow turned Whitaker pugnacious. He became involved in a bitter dispute over tithes. When the diocese sent a delegation to challenge him, he dealt with them appropriately: 'I apprehended and put them to flight.' He actually knocked several to the ground. Whitaker was powerfully built and described himself as 'a hardy son of the north'. He had striking pale eyes, almost green. When he smiled, he exposed a set of false teeth made from ebony. 'Ivory,' he explained, 'would ill-suit the gravity of an antiquary.'

For all the bluster, other glimpses of John Whitaker in Cornwall suggest someone a little softer. In time, he won over many in the parish. A letter he wrote in August 1796 explains how reaping had kept him all week from his books: 'in the fields from 9 to 7 and on Saturday evening . . . had all the work folks into the kitchen at night, to dance, drink and sing.' Soon the pews of St Rumon's were filling up: '67 families,' he recorded at this time. 'No dissenters.' When he decided to get married, he spent a month approaching 'fifty fair maids' before settling on one Jane Tregenna. They had three daughters and Whitaker proved a good father. He liked children, 'to whose level,' observed one friend, 'he loved to descend'.

To begin with, Whitaker confined his literary work in Cornwall to the old arena. He sent pieces up to London for publication in journals that shared his leanings – the *English Review, British Critic,* and *Anti-Jacobin* – and used the money to buy books for larger historical projects. His pieces were not the sort of pointless articles that saw 'both sides': *Mary Queen of Scots Vindicated, The Real Origin of Government, The Course of Hannibal over the Alps Ascertained, Origin of Arianism Disclosed.* British and world history, current affairs, all had sorry gaps in their study that needed filling with his theories – on the sophistication of pre-Roman Brits, the Jewish origins of early heresy, and the catastrophes that would result from the 'liberty' being pursued in France and America. He also began to accrue ideas about subjects closer to his new home. His two-volume *The Ancient Cathedral of Cornwall* is a platform for his own grand ideas about the early Church, Druids and Gothic architecture.

At the same time he found himself drawn into an area of research which did not allow for such polemic. He began to assemble material for a *Parochial History of Cornwall.* When it came to his own parish of Ruan Lanihorne, he found it grew and grew, absorbing him so completely that for a while the book took a back seat. His 'History of Ruan Lanihorne' was something new, and ahead of its time. Its originality was not lost on Whitaker: 'I have thus with a hasty hand sketched out a draught of a Parochial History in a single parish . . . It is the first draught I ever saw of the kind.'

By now, the last decades of the eighteenth century, the popular appetite for British antiquity had run out of sources. The beginning of archaeology was still several decades off. Classical writers had already been plundered for every little hint of what they might reveal of ancient Britain. Geoffrey of Monmouth's British History was old hat. To sift through Gildas and Bede and the *Anglo-Saxon Chronicle*, with their repetitions and speculations and invention, left precious little that was solid. Antiquarians saw only Druids when they looked at old stones. Dr Johnson claimed that: 'All that is really *known* of the ancient state of Britain is contained in a few pages.' But, he scoffed, that didn't discourage the growing tribe of historians; he singled out Whitaker's own early work as an example: 'Yet what large books we have upon it, the whole of which, excepting such parts as are taken from those old writers, *is all a dream*, such as *Whitaker's Manchester*.'

But here, in the couple of miles around his rectory, Whitaker discovered a fresh way of revealing the past: through old walls and rubble piles, ruins, fields, oral history and toponymy. He learned (or partially learned) to pick apart the compound Cornish of the area's farm names and field names (Treworgy – *tre war guy*, 'the house on the water'; Trustein – *trev is yn*, 'the house below the hill'). In discussing Treviles ('lower house'), he noted the prevalence of 'lower' (*-viles, -gollas*) in place names and suggests that, to escape the winds, the Cornish 'clung to the foot of hills, and dived into the hollows of the earth'. He dubbed his pioneering toponymic studies 'the process of critical chemistry', a phrase that hints at the magic of alchemy, extracting something of value from the base earth. Whitaker was lucky to have a ruined castle in the middle of his parish. By combining John Leland's brief record of it with local tradition and a few documents, and by examination of the ruins themselves, Whitaker pieced together its vanished magnificence: the main tower fifteen metres high, six other towers and a water-gate opening out on to the river; the Great Hall and the dungeon beneath, lightless and slopping with each flood tide. In the surviving ruins, Whitaker found the remains of cavernous ovens and a brewhouse, and outside the castle walls, evidence of huts and stores and cottages and a number of other opulent homes. He collected folk memories, many generations old, of riotous feast-days at Ruan, when Cornish landowners brought teams of wrestlers to compete in the waterside ring, of hurling matches when the 'lowlanders' of the parish would try and get a silver ball up to the 'lone house at Treworga' while the 'highlanders' aimed to throw it through a broken pane in the church. Whitaker not only reconstructed the original plan of the castle and the shape of the small town around it – he restored life to them.

At its height, in the fourteenth and fifteenth centuries, the castle was the seat of the Archdeacon family. Rents from their lands – from as far afield as Devon – were converted into masonry and turrets and towers. But in the 1500s the Archdeacons came up against a problem: a generation of daughters who married and moved away. The family rents were swallowed up in other estates; Ruan castle fell into disrepair. The houses around it were abandoned. Their stones were taken for stock-walls and piggeries. The towers of the old fortress, wrote Whitaker, were 'left for the daws'. The port remained, for a while. Briefly the silting of the quays upstream made Ruan busier. But just a few decades before Whitaker arrived, visiting ships were forced downstream again, to Kiln Point, and he heard the complaints of the Welsh coal-traders whose sloops were always going aground. 'Such an evident declension,' he wrote, somewhat clumsily, 'have we here in the navigableness of Ruan Creek!'

Whitaker's manuscript was never published. It still sits in the Courtney Library archive in Truro, in his own rough card cover and sewn binding. Yet looked at now, it appears prescient. His innovative ways of finding the past in the land anticipate not only the development of archaeology in the nineteenth century but, more recently, the entire discipline of landscape history pioneered by W. G. Hoskins. In the context of his own time, Whitaker was out of step. All over the country, improvements in roads, maps and coach-springs were drawing people to its more evocative fringes – not for its history, but its beauty. Since Edmund Burke's *Inquiry into the Origin of the Idea of the Sublime and the Beautiful*, published in 1759, a generation had discovered the strange notion that nature could have an aesthetic appeal. Within a few decades, Burke's text had encouraged a vogue that lasted well into the nineteenth century. People began to do what had not been done so reverently, nor in such a range of places, since the early Bronze Age: to gaze upon natural wonders.

Whereas Romantic poets were happy to contemplate landscape and let it work on their imaginations, others liked to pass judgement. In *Northanger Abbey*, Henry Tilney 'voluntarily rejected the whole city of Bath, as unworthy to make part of a landscape'. Tilney represented the growing numbers who viewed nature as cultural, as a question of taste, 'and decided on its capability of being formed into pictures'. William Gilpin's books spread the picturesque movement more widely. They each had titles like: *Observations of…* [such and such a place] *relative chiefly to Picturesque Beauty…'* From the Isle of Wight to the highlands of Scotland, from Kent to North Wales, he sifted through the nation's places and chose those worth seeing. Hundreds travelled to them to capture their loveliness with pen and paper, brushes and paints. The movement was well satirised by William Combe in his verse *Dr Syntax in Search of the Picturesque*: 'I'll *prose* it there, I'll *verse* it there / And *picturesque* it everywhere.'

For men like John Whitaker, the picturesque was degenerate, mere solipsism. Places should be examined to reveal the past, not to indulge some ephemeral yearning of the soul. Seekers after the picturesque allowed ruins in their favoured scenes as melancholy reminders of man's frailties, but for Whitaker somewhere like Ruan castle had only one use: as a reminder of Britain's noble past. His views were echoed by Uvedale Price in his 1794 *Essay on the Picturesque, as Compared with the Sublime and the Beautiful*: 'The ruins of these once magnificent edifices are the pride and boast of this island; we may well be proud of them, not merely in a picturesque point of view.'

The Fal Valley at this time, before the china-clay industry had taken off, was as picturesque a place as any in the West Country. But few visited it. Gilpin never cast his popularising eye over its sylvan views. Late in the 1790s, when Whitaker was deeply immersed in his hedge-by-hedge study of the parish, Gilpin came to Cornwall. It was his first visit and he was here to list its attractions and describe their picturesque value for the benefit of his readers. He crossed the Tamar. In Launceston, he was impressed at once by the old county town and its atmospheric castle. Heading west over Bodmin Moor, his enthusiasm began to wane. He looked at its treeless expanse and deemed it 'a coarse, naked country', a place 'in all respects as uninteresting as can be conceived'. Told he could expect more of the same if he carried on, Gilpin returned over the Tamar to Plymouth. Cornwall – perhaps the most varied and spectacular of all England's counties – never received Gilpin's stamp of approval.

I left the shadows of St Rumon's church and walked out into the sun. Just below the graveyard was the site of Whitaker's rectory. It is now a private house. At the age of seventy, knowing his time was running short, Whitaker sought to complete his earthly tasks. The years had not dulled his ambition. He planned antiquarian studies of Oxford, London and the ministry of St Neot. He would then tackle his big book on Shakespeare before moving on to put down a lifetime's thoughts on the Bible. He travelled to London to begin research. Briefly he was back in the swim of his youth. He took 'daily and nightly sallies' to libraries, went on outings. He had 'energetic and diversified conversation with literary characters'. But it proved too much. While in London he suffered a stroke, and died soon afterwards.

It is easy to make fun of John Whitaker, just as it is easy to make fun of the craze for the picturesque. But each in their different ways promoted that diligent attention to the world which makes life worth living. Whitaker himself had his admirers. Richard Polwhele, author of a seven-volume *History of Cornwall* considered it his great good fortune to have such an elevated figure in the county – and confessed that he 'walked upon stilts as the correspondent of Whitaker'. Polwhele ranked Whitaker as the greatest of the Cornish worthies, even above the polymath Humphry Davy. Whitaker's 'Intellect' and 'Talent', wrote Polwhele, 'have a brilliancy and an intensity almost overpowering'. He used the language of the Romantic poets to equate the spirit of his friend to an upland scene. In it, John Whitaker – the striding historian – becomes a part of the topography itself: 'his sentiment is as the mountain torrent, amidst shaggy precipices and romantic glens, illuminations bold and broad, and depths of shadow magnificently gloomy'.

Ruan Lanihorne is now a backwater, verdant and spongy in its silted-up creek. Photos from the early twentieth century show small boats working here, and a regular ferry from Truro, but now the stream is no wider than the length of a man. I'd managed to squeeze my boat up here last summer on the flood, with reeds brushing the gunwales on both sides, and, with an anchor hooked in the grass, we'd stepped ashore by the old castle and gone to the pub. When I'd first come here some fifteen years ago, there were still traces of the castle wall in an agricultural yard crammed with clobber. But the yard had since been cleared, and a group of flats and maisonettes built on the site of the castle. In six years not one of those properties had sold. On the grass outside stood a line of estate agents' signs, thrust into the ground like banners at a medieval tournament.

*

Out of Ruan Lanihorne, a lane led up the hill and a smaller one branched off it, high-banked and tunnelled with trees. The sun flickered through them as I walked. At the top of the hill, I picked up a bridleway which dropped to the marshy top of Trelonk Creek. Arms raised through a crowd of nettles and wild carrot, I watched the water reach my ankles, then my calves. The footbridge was partly submerged and I felt with my boots for its boards. In Ogilby's *Britannia* of 1675 – the first road atlas in Western Europe – is a page entitled 'The Continuation of the Road from London to Land's End'. Head west from Piccadilly across southern England, and this is the route you would have taken: this bridleway, this bog, this stream-crossing.

It was evening; I was nearly home. Everything was becoming more familiar. There against the hillside was our neighbours' herd of Charolais, set into the pasture like figures carved on a chalk hillside. A gap in the hedge revealed a distant slice of the estuary: the tide was almost on the mudflats. Coming in or going out? I found myself calculating: *Two days since full moon . . . so, springs . . . early evening: the first of the coming tide . . .*

On the wall at the top of our track, serving as postboxes, were two lengths of earthenware sewer-pipe. From one, I slid out a copy of the parish magazine. Its back page was doodled with snail-nibblings. I flicked through it as I walked. The Tregony Seniors Club would be entertained by the Cober Valley accordion band, the Roseland Quilters were to meet and there was to be a fund-raiser in aid of the Cornwall Cuddle Centre.

It was all getting very local. At the abandoned barn, I looked for the owl high in the ivy-covered gable, and saw its back facing out and its folded primaries the colour of chamois and ash. In the twilight, I pushed on down the grass-centred track and along the low ridge where the land dropped away to the creek below. Across the river the oaks were still brown and unleafed, and at Penkevil Point seven white dots in the lower branches were egrets. Rising from the arc of the field ahead were the tops of the yews and then the brick chimney stacks and the slate roof and windows of our house; and there was Charlotte digging the raised beds and the children bouncing on the trampoline, now tumbling from it, scrambling over the fence, running up the last few metres of the track with outstretched arms to greet me.

# 11 | RUAN CREEK

Ruan: after Rumon, Cornish saint; Creek: via both *krik*, Swedish dialectal, 'corner, bend, creek', and *crique*, Norman; a name likely to have been used less by local landsmen than the seaborne invaders who pushed into such tidal inlets.

DAWN, AND THE WIND IS blowing hard from the south. Through the bedroom window, I can hear it in the beech leaves with a sound like the sea. Not quite the sea. It lacks the crazed hurl of waves breaking; it's more fluent, more measured. I slide out of bed and pull apart the curtain and look across the field to the creek: the water is streaked with breeze. Everything is damp and grey but it isn't raining. I dress, pull on boots and go out to look. Or rather, to inspect: the hawthorn hedge I tried to lay in February; the beech-whips we planted beside our single patch of lawn; the old crab apple and cherry and the pear tree; the Kea plums I pruned so savagely; the fifty ash trees we put in and the patches of bare ground we seeded. Everything looks different when you're responsible for it; the carefree gaze of the walker was gone.

Over the next few days, I settled back into a familiar rhythm – early to my desk, and afternoons filled with outside jobs. Cracked panes in the greenhouse, broken slates, dead elms that needed felling, piles of rubble to shift, all the loose ends of the building project, paperwork, invoices. I began sanding the gunwales of my boat. I took the billhook to the nettles which crowded all the edges, and slashed back brambles. Growth wasn't quick enough where it was wanted, much too quick where it wasn't.

The shelduck were gathering; their liveried figures were scattered over the mud at low water, whimbrel too, but the curlew had left, as had the redshank and wigeon. Even the Canada geese were scarce. Instead the hedges and fields were full of the sound of finches and chiff-chaffs, and a green woodpecker which I kept hearing but could never quite see, and cock pheasants' woody clucking as they jousted, and partridges in pairs, and squeaking swallows swooping low over the field on their way north.

But no swifts. During the building work we'd had to repack the gable-top where the chimney threatened to topple over and, although we'd made sure to leave an opening, I was thinking perhaps it wasn't enough. They were late. Thinking of last year's swifts – two, sometimes three – chasing each other around the house like fighter jets, and the Doppler effect of their screeching as they raced overhead, suddenly brought back the first couple of months here: the hard-to-believe sense of arrival after waiting so long, and the place itself, as we found it. The swifts' absence left the feeling that our tampering had somehow emptied the house of its spirits.

Some weeks ago, we'd ordered a piece of cargo netting for a hammock and now, one evening, we took it down across the field to select a tree to hang it in. Around the creek were a number of pedunculate oaks that present something of a puzzle. They'd sprung up centuries ago, in decent soil, in good faith. But since then, the shaley cliff had been undercut so that the great bulk of each tree now hung in space. Their weight was supported only by a small section of roots reaching back into the shale. It was possible to climb into these trees from the field side and in one, two boughs diverged, then grew towards each other again. Between them we lashed the netting. Lying in it, dropping sticks into the water below and reading stories with the children until it was dark, gave the sense of a double suspension, by the net and by the balancing act of the tree.

Already there was the question of firewood – even now, in spring. I still had a large pile from the trees we'd dropped the year before. Once split, I estimated they would see us through until the new year. What I could scavenge and coppice now would be about ready by then. From the field boundary, I chose a couple of elms, cut them down and cleaned away the brush. I lopped from an old beech a bough that was threatening to topple it. I spent hours with axe and wedges and saw, easing into that particular contemplative state of the log-chopper. With a couple of friends, I took the boat up a side creek and we towed back a few fallen elm lengths. I corded and stacked them and now, with several such piles around the place, I gave each a smug glance as I passed their end-on neatness and the cavities for crawling creatures and summer bugs: 'Every man,' wrote Henry David Thoreau in *Walden*, 'looks at his wood-pile with a kind of affection.'

A century after Thoreau, the Wisconsin conservationist Aldo Leopold left his own reflections on the rewards of wood-chopping: 'If one has cut, split, hauled and piled his own good oak, and let his mind work the while, he will remember much about where the heat comes from.' Looking into the flames of his grate, Leopold recalls the lightning-toppled oak that provided the logs. He remembers his 'saw biting its way through the rings, stroke by stroke, decade by decade, into the chronology of a lifetime, written in concentric circles of good oak'. He cuts first through the few years he has owned the farm, then to the time of his bootlegging predecessor and the 'dust-bowl drouths' of the 1930s, sawing through the year of the abolition of state forests in 1915, the sawfly epidemic of 1910, the setting up of a forest commission, the years of forest fires, 'over-wheating', soil exhaustion, extinction of the Wisconsin turkey and the elk, back to 1865, the 'pith-year' of the tree, when John Muir tried to buy a farm thirty miles to the east in order to create 'a sanctuary for the wildflowers that had gladdened his youth'. As Leopold saws back and forth through the oak, the past drops to the ground to form a brown ridge – 'called sawdust by woodsman and archives by historians'.

We were stone-facing my studio (aka 'the writer's block') and whenever I could, and in an amateurish sort of way, I put in a few hours with the masons, Ash and Ben. It was like a jigsaw, but one in which each piece had to be shaped to fit. I'd found the stone at the

Callywith quarry near Bodmin, the closest match to the killas of the house. It had a good natural face and was a little more orange, more ferrous than the rock from the creek, but would, I thought, weather down in time. It had been dumped around the building in three large piles and working it meant a lot of rummaging, turning each stone over to see how to shape it and how the grain lay. Then if you tapped it with the scutch hammer, it would drop open along the sediment layers.

The same easy cleaving happened when you struck a log right, and I found myself drawing parallels – the annual growth that formed the wood, the floodings for the slate. Each was built up year by year, the wood by the mysterious forces of cell division, the stone by gravity. Wood cuts most easily *across* its age-rings, against time, while the stone breaks in the way it was made. It is as if the binding of living things must be stronger than time, so brief is their existence. Stone, on the other hand, grows not in years and decades, but in thousands of years. It has no urgency.

Our children had been given a picture book called *A Street through Time: A 12,000-year Journey along the Same Street*. On the first page is a Mesolithic settlement, on the next a Neolithic one, and so on. Each time you turn the page there is the same topography, the same river and the same hill – but everything else changes. The forest is cut back, huts appear and disappear, defences come and go; early on, a barrow and a stone circle pop up on the hill, fall out of use and are swallowed up again by the forest; an Iron Age fort becomes a Roman fort among whose ruins a medieval castle is built which is burned during the Civil War and whose broken walls, on the final page, serve as a visitor attraction. The foreground of that last page is a frenetic street scene with cars on the road, planes in the air, wine bars in basements, pedestrians on mobile phones and on the river, a small dredger and a couple in a rowing boat.

I imagined a version of that narrative here, on the tidal section of the Upper Fal. The first pages would be similar: the shoreside attracting early settlers, the tumulus in our field, the Roman garrison up-river at Golden Mill, the castles at Ruan Lanihorne and Tregony, the church at Lamorran. On the late medieval page, in our little creek, a couple of punts are pulled up on the shingle. A lugger is bringing in maunds of mussels. A small settlement stands on the shore, a scatter of barns and cottages and stock-pens ringed by the enclosures of the demesne lands. The stream drives a small mill. At the centre of the scene is the medieval manor – mullioned windows and high chimneys, and beside it the chapel. The next page is the late eighteenth century: the manor house in poor repair, only partly inhabited, but a new quay on the creek, with a lane running down to it and several people disembarking from a small coaster. Bales of goods lie on the shingle. In the woods across the river they are felling oak, stripping the bark for the newly expanded tannery.

It is at this point that our version diverges from the steady modernising in the picture book. The next page here, the mid-nineteenth century, would show the buildings in ruin, and only a much smaller house, our house, among the old walls. Otherwise, the scene is emptying of people, emptying of river traffic. The silting process is beginning to accelerate. On the following pages there are fewer and smaller coasters, then none at all. The quays fall into disrepair, the lime-kilns disappear. The track is less used. There is some activity in the woods, where timber is still taken for the tannery in Grampound, but that comes to an end in the later twentieth century. On the last page are just trees and mud flats, and a cluster of roofs in the wooded emptiness of the valley.

There's another possible version of the final pages, separated from what is here now by nothing more than the passing opinions of a few landowners. In the Cornwall Record Office, I came across a large set of maps dated 1832 and entitled 'The Fal Valley Railway'. The planned route of the railway is marked with a bold red line, curling down from the clay pits, following the meanders with its own elegant, swan's-neck curves. One proposed terminus was here at Ardevora. Had it gone ahead, the latter pages of the story might look a little more like those in *A Street through Time*, a little like this:

> In the mid-nineteenth century, with the clay industry booming, the railway brings wagons. Wharves are built for lading ships and each year, as the river clogs, so the silt is scooped out and barged down to the open sea. Dwellings spring up around the wharves for stevedores and shipwrights, who turn out working punts and fix sails and spars. From the late nineteenth century, small villas are added to the functional buildings. The last page has a two-lane road where our track now runs and a busy settlement at the creek with a car park and gastro-pub, and small yachts are moored fore-and-aft in the side creek...

In the end, a few of those who owned land beside the river dragged their feet, and the railway never happened. In history, as in evolution, it is tiny events that determine the shape of the world.

Despite its current isolation, I am always surprised by how many people have a connection with Ardevora. Several of those I sail with in the fleet of Falmouth Working Boats, the old oyster dredgers, have worked up here – renting the grass-keep, installing the old pump, putting on the roof. Tony the plumber came round to talk about the bill for the pipework. 'Just there – ' he pointed at the wall in the kitchen – 'someone's dinner was thrown against the wall.' His father had lived in a tied cottage in the village and had worked the fields down here by the river. As a child, Tony had sat up on the tractor with him and sometimes been allowed to drive back to the stack-yard. Later he worked down here for 3/6 a day. He remembered the farmer, Mike Rundle. 'Always with a gun under his arm.' But, try as he might, he couldn't quite recall whose dinner had been spattered on the wall, nor why.

One of the stockmen from the farm said he had a friend who knew the house and could he bring her round? Averil was one of Mike Rundle's younger daughters. She hadn't seen the house in thirty years and, heaving herself from the car, stood for a moment just to look.

'Yes,' she whispered. 'That's it.'

Her family had come to the farm in 1932, when she was only six months old. She lived here right through the war and beyond. She was number eight of ten children and her father, Mike Rundle, was a rabbit-catcher. He began here as a tenant, but rabbits must have been plentiful because soon he had caught enough of them to buy the house. It was Mike's family who sold it to David, and David sold it to us. David was the one who put in the electricity. He was a surveyor, but when he retired he kept pigs, and we were still finding evidence of it: papers relating to their sale and slaughter, the photo of a queen of a pig – a giant sow – and, like the totems of an ancient cult, little ornamental clay pigs half-buried in the undergrowth.

I was seeding areas around the house – hacking back brambles, raking the bare earth and casting permanent-pasture seed-mix to try and get the grass up before nettles and hogweed took over. Barrow after barrow I filled with stones and old roots and random pieces of buried metalwork, then tipped it out and raked it in along the hedges. I kept thinking of the Fal River Railway – not so much how everything here might have looked if it had been built (how big would the settlement be by now?), but the sheer slog involved in making it.

I remembered from somewhere a picture of the construction of the London and Birmingham Railway. It was June 1837, a few years after the proposed railway here, and the trench for the Tring cutting was twenty-four metres deep. Shrinking away far into the distance, at twenty-metre intervals, were rigs built to haul up the spoil. Down at the bottom, every piece of ground was removed by hand, shovel-load by shovel-load. The 1830s began an age of digging that altered the shape of the land more dramatically than anything since the rapid sea-level rise during the Mesolithic. Within the space of two decades, six thousand miles of railway track – longer than the entire coastline – had been laid across Britain, on hand-dug embankments, through hand-dug cuttings and tunnels. And then there were the hand-dug canals, the wharves and roads.

Extreme digging had long been a habit of antiquarians. From the mid-eighteenth century, and with increasing zeal into the nineteenth, learned men led parties of shovellers out on to chalky downland, into the hills, anywhere there were barrows or tumuli to hack into. Over twenty years, from 1757, the Reverend Bryan Fausset managed to open seven hundred and seventy-seven barrows around Kent, including on one occasion nine in a two-hour period. In the early nineteenth century, William Cunnington and his patron Sir Richard Colt Hoare opened four hundred and sixty-five barrows around Wiltshire. Although the methods were crude and shovel-blades destroyed a great many relics, most of the early diggers at least attempted to record their finds. The Reverend John Skinner – the same Skinner who left the most detailed early account of Aveline's Hole – opened twenty-six barrows in Somerset between 1815 and 1818; records of his digging run to ninety-eight manuscript volumes.

The thought of buried treasure was never far away. Fausset discovered a disc-brooch decorated with gold, garnets and turquoise.

Skinner found a collection of amber beads at Priddy Nine Barrows (a couple of miles from Aveline's Hole), and encased in one of the beads was a bee. When he proudly showed it to a friend, a 'fair lady', she dropped the bead and it shattered. Skinner wrote a poem about it, which included the lines: 'Escaped at length our Druid's bee / Fulfills his mystic destiny / Which dooms whatever is to be.'

In 1838, the Reverend Charles Woolls wrote an anonymous skit, having witnessed the opening of the Shapwick Barrow in Dorset. A parody of Hamlet's conversation with a gravedigger, Woolls's work features the 'Song of the First Barrow Digger':

> Clasps, Celts and Arrow-heads, I'll try
> To Claw within my Clutch
> And if a shield I should espy,
> I'll vow there ne'er was such.

Just a year earlier, in 1837, a group of quarry workers on Bodmin Moor had struck lucky. Not far from the Cheesewring, they opened a chambered cairn and found a spectacular early Bronze Age cup of ribbed gold. Such a find could not be kept quiet for long and it passed through steadily more important hands, until it reached those of the new queen, Victoria. Then it disappeared. Only a century later did the cup show up again, when her grandson George V was found to be using it to keep his collar studs in. It is now in the British Museum.

Field archaeology remained a pursuit for enthusiasts rather than professionals. They were amateurs in the literal sense – driven by love for the past and the land which revealed it. Sabine Baring-Gould was one of these proto-archaeologists, a habitual tramper of the western moors. As late as the 1890s, he carried out Cornwall's first dig above Trewortha March on Bodmin Moor.

Like so many antiquarians, Sabine Baring-Gould was not only a parson but a man of titanic energy. He had fifteen children, served his parish with diligence, restored both his family house in Devon and the church nearby, and designed and built cottages for his own tenants. He died in his ninetieth year. When not exploring antiquities, visiting parishioners, improving the lives of his tenants, or collecting popular songs and folk tales, he spent his days standing at a raised desk in his study, writing. At one point he had more entries in the British Library catalogue than any other English author. He wrote forty novels (many in several volumes), numerous books of short stories, hymns (including 'Onward, Christian Soldiers'), sermons, songs, poems and an opera. In more than a hundred books of non-fiction, he covered such subjects as werewolves and trolls, the origins of religious belief, the byways of Iceland, the virtues of patience, the life stories of Napoleon, Nero, the Caesars and three thousand six hundred saints (in sixteen volumes).

But of all his pursuits, striding through remote places looking for ruins was his favourite. Archaeological historian Malcolm Todd described his work on the moorland antiquities of the south-west as 'his love, his obsession'. Baring-Gould had a fascination not only for the way the land held the past, but for the pure spirit of place. In his *Book of the West* he quotes Scotus's notion of 'seity' which suggests that each person has 'something peculiar to himself, which makes him different from all other persons in the world.' What 'the learned Scotus said of individuals,' proposes Baring-Gould, 'may as truly be said of localities.' His best novel, *Mehelah,* is a haunting Brontësque drama defined above all by its setting – heavy with mud, pale with mists – in the Essex marshes. Charles Causley singled out from Baring-Gould's various literary merits a 'feeling for the peculiar ambience of a place…' and 'the ability to convey…the sum and substance of a particular locale'.

Two episodes from a family trip to south-west France, picked out years later by Baring-Gould himself, help to explain his love of the past and of place. It was 1850, and young Sabine was out botanising when he spotted some pieces of mosaic in the grass. 'Oh!' said the farmer. 'There are whole pictures in the ground in beautiful colours.' Organising an excavation himself, Sabine discovered a sizeable Roman villa, and one of such importance that the *Illustrated London News* ran an entire feature on it. He himself wrote the article. He was sixteen.

The other incident was both more fleeting and more telling. Baring-Gould was one of those capable of feeling sudden ecstasy at the mere sight of landscape, and it was this faculty as much as anything that drove his years of exploring. One day he was in a carriage coming up a hill when, over the brow, he spotted the distant peaks of the Pyrenees. Each of the summits was 'turquoise-blue tipped, streaked with silver'. It was, he wrote, one of those 'moments unforgettable in men's lives, moments of bursting, overflowing joy'.

One evening at the end of that week, we all took the boat out and sailed down to where the Truro river joins the Fal. Ever since coming to Ardevora, we'd been wanting to visit a small creek a couple of miles upstream from here. The day had been close and sultry, but now the cloud had broken up. There was little wind. I rowed the last stretch into the creek and shipped the oars. Oak woods pushed down to the water's edge; fallen trees lay half-submerged in the tide. To one side were platforms of salt marsh. We drifted on in silence, and saw a white house on the creek's edge.

For a number of years, all through the war, Charlotte's mother and her grandmother had lived in that white house. Charlotte had visited it as a girl, but remembers little. Her mother had lived in a number of places when young; she had brought up her own family near Salisbury, but whenever she spoke of her childhood, it was always here, in this creek. When Charlotte was seventeen, her mother died and, since moving here, to the upper reaches of the Fal basin, she'd been trying to remember all the stories her mother had told her, looking again and again at the few tiny photographs that remained. In the long months trying to sell our old house, I had always felt it was me that could not let go of living up here. But now we both wondered – was it some strange pattern of inherited memory that had led Charlotte to live in such a strikingly similar place as her mother?

I remembered a story Pete Herring told me. His father was from Sussex originally but moved to Yorkshire where he married a local girl and became an itinerant herdsman. When Pete was still quite young, his parents came into a little money and decided to buy a farm. Land was cheap in the south-west and his father took a train to hunt in Cornwall, in the area around Redruth and Camborne. But he found nothing suitable. With the money still in his pocket, he started back home. Changing trains in Plymouth, he happened to see an advertisement on the platform for a small farm a little way up the Tamar. He went to see it at once, and bought it. Thirty years passed. Pete's sister married a man with an interest in genealogy and they were all astonished to find not only that there used to be Herrings living in the area, but that their branch of the family was descended directly from them.

A day or so later, Charlotte and the children left for an extended tour of relations and friends up-country. I prepared to leave myself, to head west towards Land's End. But first there were jobs to complete – finishing off laying a path, some drain business, ordering more stone, and replacing the shock-cord clips on the boat's cover. By the second evening, everything was done. I went down to the creek to sleep in the cargo net. Laying in the darkness listening to the water right below, I could hear the plop of mullet. The tree above contained me. Much later I woke and the moon was sliding through its mesh branches. Then it was light and the geese were returning from the fields. Around me, the rising sun made olive-green walls of the boughs and canopy.

Walking back across the field a little later, I was suddenly aware of a distant sound. I looked up. It took a moment before I spotted it, black against the sky – a swift! It headed for the house, banked into the gap between the yew trees and the chimney, and was gone. It reappeared almost at once, climbed high and then went round again. As I watched, two others joined it, dashing round and round, with their high squeak right over my head. Then, as they passed the gable, one of them swooped in under the bargeboard. They'd already found the old nesting-site. After all the months of stripping and banging and improving, the house was alive again.

# 12 | TOLVERNE

Cornish *Tal-*, 'brow, front'; *bren*, 'hill'. The softening of the 'b' sound to 'v' is probably caused by a lost definite article.

It was just after seven when I pulled the pump-room door closed and crossed the field, heading for Cornwall's far west. The morning grew warm and damp; low cloud tangled in the treetops. At the Old Quay I took off my boots, rolled up my trousers and stepped down into the creek. The stream trickled through its deep gully and I followed it for a while, feeling the cockle-shells sharp beneath my feet. Down here, the top of the mud banks were at eye level. A lone shelduck stood on Stick Island, like a commuter who'd missed the last train. On the far side, I hauled myself up the oak roots, retied my boots and set off along the ride.

It was wonderful to walk in the woods that morning, following the creek-edge, with everything muffled and dripping, and the familiar tracks giving way to places I'd only peered at from the boat. The drizzly rain coated everything, oozing and running down matted foliage. For an hour or two, I followed deer paths through the understorey, weaving between squat holly and ivy tendrils, crossing glades of bluebell and ground elder. Trees lay where they fell, half-rotted in the mulchy ground or flopped on to others, the violence of impact caught in the ripped-off limbs. Where the woods gave way to the creek, oak boughs were part-entombed in the mud, like the fossilising remains of extinct megafauna. A week ago, I'd been down on the creek when I heard a noise, a deep rumble and shearing that went on just long enough to make me think it was something catastrophic. Then there was a final swish and I saw the canopy of an oak falling and coming to rest at the water's edge. It was a day of no wind; I was mystified by what caused it.

At midday I left the riverside and followed a steep valley up to

Tolverne Barton and into one of those forgotten, time-rinsed corners with which Cornwall rewards path-strayers and the persistently nosy. From the early fifteenth century, the house had belonged to a minor branch of the Arundel family, and they lived there in modest prosperity for several generations. Then came the heady maritime years of Queen Elizabeth, and Sir John Arundel of Tolverne put his sea-skills to good use. It was he who sailed the first ship back from the 1585 exploratory journey to Virginia. He continued his adventures but in a quest for the mythical island called Old Brazil, he lost all his money.

Now Tolverne Barton was a backwater. No one I knew had seen it. My requests to visit the elderly brother and sister who rented it had been refused. Taking the direct approach that morning, I strode into the farmyard. There was no one about. I walked towards the house. Its ground floor was half-hidden by un-pruned shrubs and I could see missing panes in the mullioned windows.

'Hey!'

A man was approaching in orange overalls. I explained myself.

'They're not here any more,' he said. 'My aunt's gone to a nursing home, and my uncle's just moved to a bungalow.'

It was not hard to see why. The house looked barely habitable. Beside the doorway a lean-to roof had collapsed; in the corner, an upper window was so completely covered with ivy it looked as though the building was wearing an eye-patch. We tramped down waist-high hog-weed to reach the front of the house.

'Needs a bit of work. I'm supposed to be moving in shortly.' He clearly wasn't looking forward to it.

Parting the grass, we looked into an arched portico. Green mould covered the granite steps and pieces of broken wood lay inside in the dust. 'Under there – ' he pointed at the damp floor – 'is a tunnel, runs all the way down to the river. And there – ' he stepped back to point up at a broken window – 'that's where Henry VIII stayed.'

The places that Henry VIII was supposed to have stayed in Cornwall are certainly fewer than the stories of secret smuggling tunnels, but each claim has grown from the sheer pleasure it gives in the telling. No record survives of Henry VIII even reaching the Tamar.

Following Sir John's reckless island-hunting, the Arundels were forced to sell the manor at Tolverne. The family slipped from view. Some moved away but a number stayed on in the parish where they'd been born. Generation by generation, they slid deeper into its backwoods, known only to neighbours, dropping down the social scale from gentry to yeoman, from yeoman to farm labourer. Since we'd arrived here, I'd often wondered whether spoken use had worn away the first syllable of Arundel like the sharp edge of a pebble, and whether Mike Rundle, rabbit-catcher of Ardevora, might in fact be the descendant of Sir John Arundel, Elizabethan explorer. According to an elderly member of the Rundles I met, that's exactly who they are.

Below Tolverne is a field and thick woods that drop away to the river. Leaving the house, I entered the woods and stood amidst the tall coppice oaks, tilting my head back to watch their barley-sugar shapes rise towards the now-clear sky – until a breeze dumped the leaves' moisture on my face. Through the trees, I could hear the sound of a very large engine. A few days before, I'd seen from the ferry the newly moored, thirteen-thousand-ton specialist-reefer *Summer Bay*. This stretch of the Fal is so deep and narrow that large ships can lay up for long periods, hidden like the plundered craft of old, invisible to all except those who come down to the car ferry and spot in the woods the improbable shape of steel super-structures, gantries and funnels.

It is a crossing-point of great antiquity, but for the last few centuries has gone by the name of King Harry – King Harry Reach and

King Harry Ferry. One theory links the name to a larger-than-life boatman called Harry, another to the pleasing notion of King Henry VIII and Anne Boleyn using it on their honeymoon (after a night of Tudor passion at Tolverne), with the additional possibility that Henry swam across the river with the Queen on his back.

But here in the woods was something more convincing – a ruined chapel, dedicated not to Henry VIII but to Henry VI. It took a while, but in the end I found it – smothered in low holly, cushioned by leaf mulch and dark as dusk in the shadows. It was no more than a trace of perimeter wall, signs of flattened ground, some dig-and-fill terraces and, leading to it through the trees, the shallow U of a disused track.

The link between the chapel and Henry was established by the historian Charles Henderson in 1923. He too came to this remote wood and found the 'small enclosure containing some ancient and stunted trees'. He then matched the site to a will in the Prerogative Court of Canterbury of a 'certain Reginald Wolvedon' that granted five shillings to the 'chapel and store of Our Lady and King Henry' at Tolverne. The will was proved in 1528, at a time when the regard for Henry's holiness was at its height. Ever since he'd been murdered in the Tower in 1471, Henry's name had been invoked by a growing band of the troubled and afflicted, who managed thus to revive the dead, avert suicides and cure plagues. Henry's name helped a burnt baby in Rutland to grow a new cranium, and saved a woman from death when a plum-stone stuck in her nose. In Henry's list of miracles, there is no reference to Tolverne. But he did manage to prevent a number of drownings, and once rescued a leaking ship in the Bay of Biscay. Perhaps the chapel was built by Reginald Wolvedon after some lucky escape at sea.

Henry's canonisation was well underway when it was swept aside for ever by England's split from Rome. In his hand-written

Cornish history in the Courtney Library, Charles Henderson has pasted a letter dated 23 December 1923 from His Excellency Cardinal Gasquet in Rome. Having helped Henderson's researches, the cardinal commends the discovery of the link between Henry VI and Tolverne. With a hint of five-hundred-year-old pique at England's apostasy, the cardinal reminds Henderson of the 'cultus paid to Henry VI'.

When Charles Henderson reached this damp little copse below Tolverne, he was only in his early twenties. Yet half a lifetime of exploring Cornwall was already behind him. Born in 1900, Henderson was an extraordinary figure in every way – six foot six, with radiant blue eyes and a zeal for discovery that began early and took him hundreds of miles along Cornwall's back-lanes on foot and by bicycle. However scholarly his study became – and with a hunger for ancient documents, it became very scholarly indeed – what drove him was the simple desire to see what lay over the brow of the hill, in the copse, around the next corner of the grassy lane:

> At a very early age, I began to feel a fondness for the place in which I lived. At the age of 7 I was interested in the flora of Cornwall but although I was hardly then aware of it, this interest was due to the fact that botanical expeditions gave me an excuse for exploring the countryside. Then, running through the usual gamut of boys' hobbies, I fastened upon Postmarks. It is difficult to conceive a more barren Hobby than this, but at the root of it there lay a supreme interest to me. To obtain the Postmarks of obscure Cornish Post Offices it was necessary to visit them.

Soon afterwards he was given a camera and another excuse to set off through Cornwall's remoter byways. But the usual 'snapshots of friends, boats, picnic parties, the sunny memories of dull days held no charm at all'. He decided, instead, to photograph every one of Cornwall's churches. He was ten. A year later, having processed and printed the photographs himself, he bound them into a book, its title carefully shaped in gilt letters: *Cornish Churches, Charles G. Henderson*. The following summer (aged twelve years, one month), he was on to the next thing, a project that would, by the time he was seventeen, run to four volumes and nearly fifteen hundred pages of neatly written foolscap, describing in detail the antiquities, churches and monuments of the four western hundreds of Cornwall.

A few weeks earlier, I'd spent an afternoon with just a few of Henderson's massive hoard of papers in the archives of the Courtney Library. Leafing through them, I became aware of a feeling deep in my stomach at the precocious dedication of it all. Every point of interest, every clue to the past, every quoit and barrow, every wayside cross, every round and camp and every ancient barton of west Cornwall draws his foot-borne curiosity. His method was clear. Published local histories and large-scale maps give him the bare bones of an itinerary, send him down the track or across the moor. As soon as he arrives at a site he is instantly making his own assessment of its chronology, and spotting details of his own – recording and speculating about the jut of old vaulting, the piece of reused tracery in a wall, identifying the quarry it came from. He makes sketches. He takes photographs. He talks to anyone and everyone. Nothing escapes his eager gaze. 'It was clear that he had been everywhere,' wrote one friend, 'and that his personal charm had made him a very well-known and welcome figure . . . He could see more in a day than others in a month.'

At times while reading the manuscript, with the precise descriptions, the technical vocabulary, the references, the sheer scale, I'd forget that it was not the work of some grey-haired pedant like Pevsner, but of a young adolescent with an adolescent's lurching impulses. The changing handwriting is a clue, the rounded early letters narrowing and sharpening with age, as is the sometimes arch use of expressions like 'delightful', 'beautiful' or 'bad taste'.

Among his personal papers is a more candid view of the twelve-year-old topographer. In an oilcloth notebook he wrote a truncated journal – *The Diary of C. G. Henderson and J. Zw*. Not named exactly, J. Zw appears in the text as 'Mademoiselle'. The diary opens:

> *Thurs 1 August 1912*. Started off by 2.20 train from Hayle to Gwinear. Bought 4 buns and two packets of chocolate (1/3). Took field path to Lanyon which is a very old house the ancient seat of the Lanyon family. It has a gabled porch and the Lanyon coat of arms above the front door. Took a photograph...

They go on to Gwinear church: 'a fine edifice with a tower and lesser turret. The font is old and bears representations of the Five Wounds of Our Lord. There is an old Cornish cross in the churchyard. I took a photograph of the font and two of the exterior. Mademoiselle took two exteriors and one exterior...' They follow a short cut. Mademoiselle falls and hurts herself. The expenses of the day are totted up – film, buns, chocolate, photographic developing and fare. It comes to 3/11.

Arranging his own papers ten years later, Henderson identified this day, this five-hour amble back from Gwinear to Hayle with J. Zw (she appears in a sketch, full-skirted, sitting on a gate, clutching a posy), as the moment when 'the book was begun'. He includes a photograph, taken by Mademoiselle. It is in Hayle, near the station. The Methodist church breaks the skyline. A group of three Edwardian grandees are reading notices on a wall. Henderson is the lank and indistinct figure in wide-brimmed hat and breeches. His legs are braced, as if he has been cajoled into standing still for the camera (photography, for him, was for things that don't move) – but really he can't wait to get going.

The journeys Henderson made for this early work had to be squared, in the first couple of years, with the constraints of boarding school in Somerset. When he was fifteen, illness forced him to leave. His parents sought tutors. They were lucky to find Henry Jenner, greatest of Cornwall's living antiquarians; lucky too with Canon Taylor, vicar of St Just-in-Penwith, who shared young Charles's devotion to studying the relics of the past *en plein air*. The pair were remembered hurrying through their bookwork 'so that both might go ramping together through the ancient stones of the Land's End'.

It was towards the completion of the first four volumes that Henderson's boyish roving took on more substance. He developed an interest in manuscripts and documents that not only gave a deeper context to his discoveries but set the course of his career as a historian. He revisited the Cornish churches to examine the parish registers. These in themselves he found to be 'unreliable' until set against private deeds, which he was already gathering from various sources. He started to use the British Library. In 1917 – then aged sixteen or seventeen – he discovered in nearby Chancery Lane another storehouse of antiquity: 'I ransacked the Public Record Office and made myself thoroughly acquainted with the legal records there.'

By the time he went up to New College at the age of nineteen, his name was already well known in Cornwall as a pioneering historian. The story goes that on the train to Oxford, he fell into conversation with a stranger. Chatting away, the man pointed out the paradox of New College being called 'new' when it was so ancient. In chipped Henderson: 'Like Newquay,' he said, 'in existence since at least 1480 or before.' The older man – the distinguished Cornishman Sir Robert Edgcumbe – looked more closely at his companion: 'Then you must be Mr Henderson.' On another occasion, the teenage

Henderson turned up at some remote Cornish church and asked about the age of the font. 'We're not sure,' confessed the cleric. 'You see, Henderson has not yet seen it.'

His interest in documents became all-consuming. In university holidays, he continued to tour Cornwall, burrowing into the back rooms of solicitors' offices, befriending landowners to rummage through their attics and cupboards, their dust-sheeted rooms. A collective purging of the past had gripped the country during the First World War and intensified in its aftermath. Historical records only served as a reminder of what the past had led to. In Cornwall, wrote Henderson, the old families were 'obsessed with the idea that the time had come to sweep old corners clean'. As a medievalist, he was swimming against the tide, avoiding the pull of modernism. Yet he described the fresh feeling of diving into a cache of long-neglected papers: 'In an unoriginal time when everything seemed to have been tried before, this was a delightful sensation.'

He used family connections to search Cornwall's great houses. Walking or bicycling down their drives, he discovered papers in damp byres, in tin-roofed outhouses, spattered with mouse droppings, knitted with cobwebs, hidden in the low cupboards of unvisited rooms. Sometimes he was too late. At one run-down pile, he was told the old papers had been chucked down a well. But at another, he uncovered bundles of documents in the laundry cupboard and carried them back down the drive in a wash-basket. On occasions when the owners wanted to keep the papers, he would stay up into the early hours making transcripts.

In this way Charles Henderson gathered about sixteen thousand documents: wills, covenants, leases, letters, tithe-deeds, grants, charters and bonds, pass-books, assignments, jointure statements, letters of attorney, probate of wills, marriage settlements, inventories, releases and surrenders, sales and mortgages and estate maps.

They are now stored in high-sided acid-free boxes on yards and yards of metal shelves in the archive room of the Royal Cornwall Museum. Each is tied with a length of legal pink ribbon. Loosen the bow and lift away the ribbon and the documents separate. Some are vellum pouches with the grain of the calfskin still visible. Others are streaked with damp or spotted with age. Some bear crusty medallions of red sealing wax. Open them up and in a draughtsman's hand there might be: 'This indenture Made the twenty-fourth day of June in the twenty-sixth year of the rayne of our Sovereign Lord, Charles the Second...' And reference to some parcel of pasture or wood, demesne or park of Roskrow or Tehidy or the manor of Eglosrose. Like megaliths, these documents are all that remain of past claims on the land. Invoking the King, God and anyone else thought relevant, gilded in legalese, the vellum wraps were stored against the threat of dispute, a talisman to ward off the truth that there are no tracts, not an acre or perch, that can ever really be 'owned'.

Now documents such as those recovered by Henderson are highly prized. In the last half-century, both the technology and the will to preserve the past has gathered pace. At the Courtney Library the pink ribbon – the red tape of official documents – is being replaced, for fear of dye leakage, with unbleached archival tape. The old acid-free boxes are giving way to new-generation acid-free boxes. The librarian, Angela Broome, was still not convinced: 'I don't entirely trust these new ones,' she said.

Charles Henderson gathered the documents for his own never-completed parochial history of Cornwall. It's hard to think how he might have contained it all. His notes in the archive stretch to dozens of bound volumes. Like John Leland half a millennium earlier, Henderson was saving records from the fire – not from the fanatical flames of the Reformation but from the embers of the Great War. Like John Leland, his enthusiasm for ancient documents was matched by an enthusiasm for topography. Both men sensed the truths to be found in the combination of dry documents and damp valleys, the lines of ink and the lie of the land.

Thirty years before W. G. Hoskins's *The Making of the English Landscape*, Henderson anticipated Hoskins's central idea: that the land itself is a manuscript, or many manuscripts, layered with the text of former lives. 'He had a wonderful sense of topography,' wrote his friend and fellow Cornishman A. L. Rowse. 'It was not only that the landscape came alive for him, but he could read it like a palimpsest, like a document, which it is – it told its own story to *him*.'

A. L. Rowse is one of the very few personal sources for the life of Charles Henderson. The two Cornishmen first met in Oxford in 1928 when Henderson – 'dear Charlie' – was a Fellow of Corpus Christi and Rowse of All Souls ('the first working-class man to be elected', according to his biography). Rowse was a few years younger than Henderson and, though they shared a Cornish background and an interest in its past, they were very different in temperament. In Henderson, Rowse witnessed something of the social ease and outgoing nature that he lacked: 'He certainly had plenty of fun, a time crowded with every sort of interest, sightseeing, people, places.' Rowse grew to love the lofty Henderson – 'a gangling great boy, an innocent, guileless, with a galumphing sense of humour'. And, years later, concluded that Henderson 'was one of the two or three best, unflawed, men that I have ever known'.

To the ambitious young Rowse, Cornwall was a backwater from which he had just swum free, one too parochial for serious study. But Henderson's deep immersion in its minutiae made him look at it again. With Henderson's encouragement, he began work on his own magisterial *Tudor Cornwall*. Rowse was always amazed at Henderson's knowledge. 'There are fifteen places called Penquite east of Par,' Henderson once announced. 'And at least eleven west of it called Pencoose.' When Rowse stumbled on an old property-list, he devised a little game. He would toss out the name of a farm, any farm in Cornwall, and Henderson would try and identify the parish. They might come to a place like Tregarrick and Henderson would say, 'Well, there's a Tregarrick in Laneast, and another in so-and-so, one here . . .'

'Needless to say,' recalled Rowse, 'he won.'

In one area, though, Rowse found he could offer guidance. Henderson's intellectual energies were not naturally abstract or metaphysical. 'I had to take his *general* reading in hand and told him to read the contemporary authors Virginia Woolf, D. H. Lawrence, Strachey, Eliot.'

In other ways too Charles Henderson was broadening his interests beyond Cornwall. Although he was often late back after the holidays, still sifting old papers in some remote Cornish estate, his lectures at Oxford now included other periods of European history. He became absorbed by the eighteenth century. He travelled widely in Europe. In order to read Pushkin, he learned Russian – taught by one of his own students, Isaiah Berlin. He planned a book on the decline of Church hegemony in Italy and Spain, and during summer wanderings in the Alps and through the hot byways of Southern Europe gathered details from a hundred towns and monuments. One travelling companion wrote:

> It was pleasant to see him in a new country using all the tactics that had served him in his own. He would begin on the train, reconstructing an ancient road-system from the date of the bridges as he passed; then he walked straight into the vitals of a strange town and somehow nosed out more history than was ever written of it; and, often by a happy courting of chance, fell in with the local bishop or town-clerk, who ended by asking him to stay.

He travelled with Rowse, too. They went together to the Isles of Scilly. It's clear that Rowse was a little bit in love with him, and wondered whether he might be gay. Henderson had close male friends, and there is much in his papers about his 'true and intimate friend Fred Maxse', and a coded note about Sandy Rendall. But the idea of anything physical repulsed him. 'I'd rather shoot myself.' Even so, reported Rowse, it wasn't until Henderson was well into his twenties that he 'decided to try himself out with women'. At an art class he met a girl called Cecilia, 'purchased a French letter and had her on the floor of his college room'. Rowse was devastated when Henderson married.

Later, Rowse recalled a journey he and Henderson once took together, from Oxford back to Cornwall. It was a roundabout route, over several days. Rowse noticed in Henderson something new, a curious energy. 'He was developing, flowering, restless, catching up on life, racing against time, and moving out into wider fields.' He had just learned to swim and, as they travelled west, Rowse watched him leap into the Avon at Amesbury, his pale form sliding underwater 'like an elongated trout'. Henderson did the same when they reached the Dart at Fingle Bridge in Devon. 'He wanted to try himself out physically,' Rowse wrote, 'to make up for lost time, to compensate for his bookish youth. It was as if he couldn't gulp down life quick enough.'

Charles Henderson married Isobel Munro soon afterwards, on 19 June 1933. Her father was Rector of Oxford's Lincoln College. She had been a brilliant student at Somerville and they were well suited. They went on honeymoon to Italy, to the Gargano peninsula. They swam. At Monte Sant'Angelo sul Gargano, they visited the shrine of St Michael, Cornwall's patron saint. When Charles fell ill, they returned to Rome. He wrote to his mother, and she replied at once: 'My very Darling Son . . . I am so distressed to hear of your being laid up . . . I think that bathing may have been unwise . . .'

But before her letter reached the honeymooners in Rome, Charles was dead. He had not risen for days. An infection had slowly squeezed away his breath. Isobel sat with him. Later she wrote that during those last days he had the look of an Angevin king. He was buried in the English cemetery, alongside other Britons whose flaring was brief and bright – Keats and Shelley, and the wild Cornishman Trelawney, who plucked Shelley's heart from the flames of his beach pyre.

Everyone you speak to about Henderson – at least, that limited bunch of Cornwall's enthusiasts who know about him – says this: imagine what he would have achieved if he'd lived. But what is really astonishing is to think how much, in just thirty-three years, he *did* achieve.

Among his historical archives are two boxes of personal papers. '*Unsorted*' reads a note in them. 'Oh – that's my handwriting,' said Angela Broome, remembering Henderson's sister dropping them off years earlier. 'I don't think anyone's looked at them since.'

I worked the lid free. Here was Henderson's own archive, a modest monument beside those hundreds of others he'd rescued. There were envelopes filled with letters, old school books, a cuttings file, correspondence and journals from tours of France and Germany, a stack of postcards from a trip to Austria. A letter (undated) to his mother from an early boarding school: 'it would be absolutely ripping if we could go to the beach in the holls . . . not much happening, a hockey match tomorrow'.

One of the books, a green hardback, had gold letters embossed on its cover: *Spoffkin Graphic by C. G. Henderson, 1912*, and inside: 'Is War Starting?' – a hand-written newspaper headline with a diagram below of troop build-ups. The article explained how the Spoffs of Spoffkinland were having trouble with their neighbours in Franconia and Polonia. Another report tells of an anarchist called Snojo who tossed a bomb at King Alum, and was imprisoned. Three issues later: 'Snojo escapes!' Each of the newspapers had a page or so of advertisements: 'CLICKO typewriters' or 'non-smelling tobacco'; 'Why buy ink when you can buy an ink plant?'; 'Stop that catawauling – Buy the ANTI-CAT boot filled with lead...'; 'HOBGOBLIN BOMBS – USED BY ANARCHISTS'.

I read them all, and found my face set in a grin throughout. In one way it is just the sort of spoof any twelve-year-old might indulge in in their bedroom; what is extraordinary is the detail, the extent of it, the application. Over several notebooks, the invented country is set out in its entirety. Sections explain the government of the Spoffs under Ivan, the reforms of the Spoff King Charles II, the new capital of Spoffkinville, the invasion of Spoffkinland by the kingdom of Katmandu, diagrams of medieval battles, intricate accounts of princely power struggles, the troubled relationship between crown and parliament, a multi-branched royal family-tree of the house of Livonia. In the first book is the geography of Spoffkinland – detailing the southern federal regions of Bosh, Dona Peulia and the State of Mosk, each with its own provinces listed, the chief towns and products, and a map of population density, average rainfall, cathedral cities and principal exports. There is a grammar and lexicon of Spoff – in red and white ink, explaining the conjugation of verbs, how to make plurals, rules of gender, relative pronouns, distributive and possessive adjectives – pages and pages and pages.

Baring-Gould had done something similar at the same age. In a notebook of his, found only in 2005 stuffed in a bookshelf, was a long story called *The Curious Adventures of Dr Roticher*. It featured a fantasy land deep inside the belly of a sea monster. Dr Roticher and his party were on their way to the south pole when they were swallowed up by the leviathan. Fortunately, at the bottom of its throat they slid down into a country filled with exotic plants like 'hot plum pudding tree, the roast beef tree, the fresh water bottle tree'. Crews from other wrecks were also in the beast's belly and together they built a theatre and a school and a museum. The story is illustrated with intricate architectural details of the buildings.

The young Brontës – Charlotte and Branwell, Anne and Emily – had the islands of Angria and Gondal and the Glass Town Confederacy. Their made-up places filled notebooks, grew into plays, poems and stories, and allowed such licence that to Elizabeth Gaskell they revealed 'the idea of creative power carried to the verge of insanity'. But later in their lives, all of them – the Brontës, Henderson and Baring-Gould – built literary careers on the power of place. These early fantasies were the wild spawning ground of their ideas.

Henderson's Spoffkinland is the funniest, borne of that English need to puncture with a joke anything that becomes too serious. For Henderson, at twelve, everything *was* becoming serious, his rising curiosity spreading out over the land around him. Where Spoffkinland had taken him in his imagination, Cornwall was now taking him on his long, restless legs. A later notebook has a page or two of Spoff studies, but the appeal of Spoffkinland had been overtaken by the wonders of the real world. A. L. Rowse was in no doubt about the source of Henderson's energy: 'What led him on was that he loved every hill and valley, every lane and track and field, every rock and stone of the Cornwall that no one knew, or will ever know, as he did.'

It was almost midday when I came down through the trees to King Harry Passage. The chain ferry was on the other side. I perched on a lip of root-bound soil that hung over the low cliff, and waited. The water licked at the rocks below. The river stretched away downstream, flat and smooth and windless, holding on its surface an inversion of the wooded slopes that dropped into it.

I'd known this ferry passage all my life. All those stories I heard as a child involving ferries, I would picture here – St Christopher carrying Christ or the moral dilemma my friend Zosia used to pose: of the dead man and his lover, his wife and the ferryman, and which one was to blame; the folk tale of the fox, the goose and the cabbage, and how to get them across in a one-passenger boat; and the cautionary tale of the scorpion and the horse – the scorpion stings the horse as it is carried across on its back, and as they both drown, the scorpion says: 'Sorry, it's in my nature.' I saw them all between the wooded banks at King Harry Reach.

In India such crossings are *tirthas,* among the most widespread of the sacred features of the landscape. In Hindu, Jain and Buddhist literature there are numerous references to *tirthas* – as threshold places, places between life and death, places to leave sins behind, to cross to a purer life. Deities pass to and fro between the heavens on the far shore and the mortal world on this. Before the stories was the river itself, and the pause in the journey, the waiting to cross, helped the stories grow.

The ferry here is much older than Henry VI. Henderson identified the earlier name of Kybylls or Kebellys, linking it with *ceubal,* Welsh for 'ferryboat'. The chapel built to the saintly Henry VI probably embodied a traditional sanctity that went back far beyond the Middle Ages, to a time when gods were local and presided like fiefs over sites of significance.

A faint clicking came from the chains. I sauntered down to the concrete apron where the steel links, each the size of a mango, came out of the water to their strong-points in the rock. The clicking grew louder. The ferry's ramp approached the shore and dropped, scraping up the last few feet. Ten minutes later, as the cars clanked off on the other side, I stepped ashore again. It was a crossing I knew well as a driver, but as a pedestrian, climbing the slope, it gave the strange sense of entering another land.

# 13 | PORTHLEVEN

*Porth-*: Cornish, ‘cove’ or’ harbour’; it also means ‘gateway’ and appears to have this use in the spoken and written language; only in place names does it refer to ‘harbour’, suggesting very early use. *Leven*: Cornish, ‘smooth’; Weatherhill thinks it unlikely as a direct qualifier – the place itself being so un-smooth suggests *leven* may have been the name of the stream entering the sea here.

I STAYED IN FALMOUTH THAT night and in the morning walked up through the back of the town. Its pavements and close-packed housing gave way suddenly to high-sided lanes and scattered farmsteads. Opening the map, I traced a way through the labyrinth of green-dotted footpaths, out to the west coast of the Lizard.

The route was rather more obvious on the map than on the ground. Every few minutes, I had to check it. But it was clearly very old, connecting the yards of one ancient dwelling with the next. They had names like Trengove ('the smith's farm') and Trenoweth ('new farm') and Bosahan ('house in a waterless place'). Leave the towns, the main roads and the coast, and Cornwall is filled with such places: two or three houses amid a clutter of outbuildings. The inland paths that join them are now rarely used for anything longer than a dog-walk, but they help reveal a parallel world of Cornish hamlets, or *trevs,* that date back to the post-Roman centuries. West of the Tamar there was less of the Saxon villagising that characterises much of southern England, leaving many of these settler-farms intact. Without central villages, the parishes took the names of local saints. So the eighty per cent of Cornish parishes with saints' names say less about the innate holiness of the county than the nature of early settlement patterns.

Passing through these farmsteads now was unpredictable. Many had been converted, with top-of-the range cars crouching in deep-gravelled driveways, geraniums in pots, and picture windows in the sides of done-up barns. Some were derelict, and with a few – even as you approached – it was not clear whether anyone lived there or not.

In one of the farms, I poked around in the nettle-thick yard, trying to work out how long it had been abandoned. A little further on along the track an elderly man ambled towards me with a small bag of groceries. 'That's right,' he said, stopping easily at my question. 'It's where I live.'

'Do you still farm?'

'No, no – I sold the herd, few years back now, when my wife died.'

I said something about the house looking a little uncomfortable.

'Uncomfortable?' He looked at me, as if trying to work out why that was relevant. 'It's where I live.'

He studied me. 'You trying to find somewhere? Is that it?'

'Not really. Just walking.'

I watched him go, a shuffling, baggy-coated figure winding his way back to what remained of his farm.

A few months after Charles Henderson died in 1933, T. S. Eliot returned to his native New England for the first time in seventeen years. He looked out on the once familiar ridges, the beautiful valleys rippling away into the distance, the woods ablaze with autumn colour, and he was filled with gloom. The old sheep farms and the mills were gone. 'Those New England mountains,' he explained in a lecture a couple of weeks later, 'seemed to me to give evidence of a human success so meagre and transitory as to be more desperate than the desert.'

Eliot and Henderson were, in different ways, motivated by the pervasive theme of the age: loss of tradition. The Great War had undermined faith in the past, but it was a process that had been underway for more than a century. Rapid urbanisation had dragged huge numbers of people from the places they knew. In 1935 the Hungarian-born sociologist Karl Mannheim, recently banished from Germany, warned how susceptible displaced populations were to populist ideologies, pointing to the sort of nationhood now being offered to Germans by Hitler's National Socialists.

At the same time, a group of artists was assembling in Cornwall with rather more benign ways of dealing with tradition. In doing so, they briefly placed St Ives and west Cornwall at the forefront of the modernist movement. They were drawn by the crystalline light, the bohemian lifestyle, but perhaps most of all by the region's raw topography which remained free – in their eyes – from the heaviness of the nation's past. Barbara Hepworth looked at the moors around Zennor and saw pure form, reproducing its curves and declivities in polished stone. Ben Nicholson constructed crisp geometric shapes from the harbourside houses. The Russian émigré Naum Gabo saw paintings 'in a torn piece of cloud carried away by the wind . . . in the naked stones on hills and roads . . . in the bends of the waves on the sea'. They revelled, as one critic put it, in the '*cloisonée* effect of little green fields . . . the constructional flight of the seagull'. Ben Nicholson compared striding the clifftops with his own drawing: 'Can you imagine the excitement which a line gives you when you draw it across the surface? It is like walking the country between St Ives and Zennor.' They gazed, they talked, they painted; they took the landscape and moulded it to their own ideas; they transferred it on to portable canvas and sculpted stone for collectors in faraway places. They themselves were all from elsewhere.

Or almost all. Peter Lanyon had been born just outside St Ives. For him the area was more than just shape and surface: it was belonging. When he walked the moors and saw the standing stones and burial mounds, the pump houses of the mines, he sensed the presence of his own ancestors. 'How they natter at my feet these fellows!' he wrote to Ivon Hitchens. His paintings were 'concerned with this feeling in the bones, for a county that is both very old and always fresh'. When asked once what was the chief aim of his work, he said: 'To remind people of roots.'

Lanyon became one of the more prominent of the artists. He was a colourful figure around St Ives, moving between the studio debates of his fellow artists and the quayside lives of his fellow Cornishmen. His work and his life were a ceaseless struggle between the local and the universal, between the figurative and the abstract, between place and space. Lanyon described himself as 'a place man': 'I paint places but always the Placeness of them.' By the late 1950s, he was dubbed 'the last landscape painter'. Another said that 'almost single-handed Peter Lanyon remade English landscape painting'. He himself resisted those who placed him among the other abstract artists of St Ives. 'I'm really just an old landscape painter like Constable, only they can't see it. I'll probably end up painting sheep in a field.'

The clearest impression of Peter Lanyon can be found in a series of scrapbooks combining his own notes, family photos, anecdotes by friends and reproduction of his work. Here he is in his paint-spattered studio, with the light flooding through tall windows, and here raging in a letter to a friend about his fly-by-night creative energies, the constant battle to sell his work, and here he is with a few of his five children, around the lunch-table or out on the moor, or on the cliff with the wild wind tugging at his hair. The albums have been compiled by his eldest son, Andrew. I'd phoned him a day or so ago and arranged to meet him at his house in Porthleven, but on the way I wanted to visit the scene of one of Peter's more haunting paintings.

In the mid-morning I lost the path. I doubled back, took a short cut and it ended the way it usually does – crawling through a hedge, unpicking brambles from my hair. I tumbled out of the thicket and into an open field. I brushed myself down. An old Massey tractor on the far side was topping docks. In its cab sat an elderly man in clear-rimmed glasses.

'Something rather shameful about docks in your field, don't you think?' He climbed down and pocketed the keys. 'Docks and ragwort. Awful lot of ragwort nowadays. Cup of tea?'

We walked along the lane to his house. We took off our boots in a small cobble-floored building. He pointed out the granite mullions of the windows. 'Long time ago, this used to be the house. They just left it as it was when the new one was built in the seventeenth century.' After three centuries, the fireplace was still blackened with soot.

His own family had lived in the main house since the early 1800s. It was high-ceilinged and spacious and I followed him through a hall of stern-faced portraits, across a foot-furrowed floor and out on to the terrace. We sat on a bench and drank tea. Before us was an elongated and undulating strip of lawn, laid out with a multitude of croquet hoops.

'That's three sets put together. It worked better with all those hoops.'

'It must mean long games.'

'Oh, I don't play. Just thought it looked nice.'

He kept cattle, a suckler herd of Herefords, and talked of them with the pride of a parent. 'They're wild animals – that's how I think of them. On the whole they have a happy time, looking at the sky and running around. Not a bad life really.' He paused and smiled, enjoying for a moment the thought of his beasts. Then he sighed. 'But farming – I'm too *old* for all that now.'

His conversation was full of asides, linguistic explorations. He had given classes in Cornish and was teaching himself Russian. He was immersed in Byron right now, on volume five of Leslie Marchand's edition of the poet's *Letters and Journals*. 'I don't listen to all that stuff about his profligacy. What you have to remember is this – Byron was simply a wonderful human being.'

Three horses came along the lane. The way the lawn dropped away, you could only see the horses' necks and the top half of their riders: it made them look like carnival floats.

'My uncles lived here between the wars. They were – ' he dropped his voice – 'a little *rough*...' He let the image hang for a moment before adding, without conceit, 'I, on the other hand, was rather a clever little boy.' He had been sent off to school in Devon and ended up at Oxford. Soon after his finals, he inherited the farm from his uncles and returned to Cornwall. 'And I've been here since then, really, ever since.' His voice trailed away and left an unspoken hint of what might have been, what he might have done had he not been yoked to his ancestral land.

I mentioned Charles Henderson's history of this parish.

'Oh, Henderson!' he said. 'He came to this house, of course. But you won't find more than a note about it. I rather think my uncles gave him short shrift. They didn't much care for historians. They really were *pretty rough*.'

By the time Peter Lanyon was born in 1918 his family had already left the land. In the early nineteenth century they had prospered from the mining boom in nearby Camborne and Redruth. But by the Great War many of the mines had closed. The Lanyons moved to the coast, becoming a little less spruce in the process. Peter was born into a bohemian, music-filled household in St Ives. He showed an early interest in painting and in the late 1930s was taught by Ben Nicholson. He was quick to acquire an understanding of abstraction and learned, like all the incoming artists, to apply it to the physical drama of West Penwith.

In the harbour district of St Ives, among the narrow sea-ended alleys of Upalong and Downalong, Lanyon cut an impressive figure. He was charismatic, funny and gregarious. His convictions were fiercely held and forcefully expressed. As his reputation grew further afield, so did his localism, his loyalty to Cornwall. In St Ives, he usually drank not with the other artists, but with working men in the Golden Lion, hunched in the corner in his black beret, shuffling dominoes across the table. He was mercurial. He worked in sudden bursts, between long periods of torpor. He got into fights at parties. He fell out with many of his contemporaries, notably Ben Nicholson and in an emotional letter to another artist, he wrote about Nicholson: 'TIN PAN BEN with a squeaky scream / Tried to tell the critics about CORNISH CREAM.' He made a point of pissing against the wall of Nicholson's house on his way home from the pub. 'Why don't you admit you're an abstract painter,' Roger Hilton challenged him, 'instead of all this stuff about Cornwall?' Lanyon refused. His paintings were about places, real places. They had names like *Bojewyan Farms, Trevelgan, Portreath*; they were not just images, but stories.

For him, tackling the big questions meant an understanding of place – not the regional or provincial, but the *local,* where a single field or a lone cove can conjure up a whole world. In his painting he was trying to recreate a 'mile of history in a gesture'. The past was not something dusty and ossified but raw material from which to build a picture of our own time. During the war, Lanyon served for six years as a mechanic in the RAF; migraines prevented him from flying. When a plane needed fixing, he went out to the dumps to sift through broken mechanical parts. He equated the process with his own art, the improvising, the making of one thing from the bones of another. 'My painting has developed in precisely the same way through reference to a place instead of a dump of old machinery and the subsequent building up of a separate and purposeful object.'

Charles Henderson hunted in byres and attics, Lanyon in RAF dumps and in the landscape of west Cornwall. T. S. Eliot rummaged in the storehouses of ancient myth. Each of them rescued the past in order to give life to the present:

> Old stone to new building, old timber to new fires,
> Old fires to ashes, and ashes to the earth
> Which is already flesh…

*

Some little way beyond the village of Cury, still short of the coast, I scrambled up the side of a high hedge and, in the moment before sliding down again, glimpsed an entire view. Distant hedges and fields; in the foreground a stallion, tail swishing and throat pressed against the fence of the next paddock, where two mares were grazing.

A path led on towards the coast, through the links of Mullion golf course. 'Mind your head there!' came a terse shout from the tee. I paused to watch a woman in a green visor line up her shot, alternately eyeing the ball below her and the fairway ahead, rocking on her feet before swinging. The ball soared far over the close-cropped green.

At Church Cove, it was high water. Family groups with stripy windbreaks were squeezed like refugees on to the last patch of beach. The hill of Gunwalloe rose beyond them, dropping steeply into the water.

From all of Peter Lanyon's paintings, one detail stands out sharpest for me. The picture it appears in is, loosely, of here – Gunwalloe Hill, looked at from the south. The sea is sweeping up against its slopes. Encased within the hill is a melange of shapes that dominate the painting – a standing horse, or perhaps two horses, being the most obvious. Yet the picture takes its title not from the horses but from a tiny four-legged creature on the slope above. The creature – a dog or a fox – is painted yellow and is running. The painting is called *The Yellow Runner*. So powerful is that tiny figure against the hillside that whenever I turn back to look at the whole image, it's always a surprise to find it not quite as tiny as I'd remembered.

Lanyon himself left some notes about the piece – 'Painting of a story. Runner with a message on way to stockaded horses... A home-coming.' It was 1946. During his six years in the RAF, he had served in North Africa, Palestine and Italy. Opening his eyes to the wider world, the war also created in him a yearning that never dimmed – for his home country, and for particular places on west Cornwall's rocky coast. 'Do you know what that coast means to me?' he wrote to his sister from Italy. 'It means the sea and generations of waves coming in to be broken on its shores, it means a rock like a hand in the Gulf of Bosigran and the sea pinks by the old mine houses.'

The central cartouche in *The Yellow Runner* he identified as 'the pollen and the flower, the sperm and the egg... stockade as womb'. All were representations of renewal and regeneration, ideas from James Frazer's *The Golden Bough* that had been in vogue before the war. In other pictures, Lanyon used horses to represent ancestral connection to the land. But in that yellow animal – legs stretched out fore and aft, dashing across a bare hill with its unstated message – is something more enigmatic.

Peter Lanyon himself was a great enthusiast for speed. If he hadn't been a painter, he said, he'd have been a racing driver. In one of the scrapbooks Andrew recalls them driving so fast along the narrow roads of Penwith that sometimes, after a bump, there was a moment of quiet and weightlessness as the car became airborne. But Lanyon was not simply after exhilaration. Driving fast was part of his life-long quest to re-jig the view, to see it afresh. 'Peter's perception of the landscape from his car,' wrote one critic, 'is one of his unique contributions to landscape painting in this century.'

Lanyon, it was said, was the only one of the St Ives artists who knew where the wind was coming from. While the others looked on with the eyes of newcomers, Lanyon had to re-invent the familiar. He needed to create 'a disturbance of equilibrium'. One photograph shows him bending over to look at a landscape backwards through his legs. He snorkelled. He lay on the edge of high cliffs, climbed into old mines, always looking. A sense of danger and unease became the required stimulus not so much for the work itself as for those moments in which it was conceived. 'Without the urgency of the cliff-face or of the air which I meet alone, I am impotent.'

In the late 1950s Lanyon discovered gliding. At Perranporth, the Cornish Gliding Club made use of a disused wartime airfield, cliffs ninety metres high, and the westerly wind to provide 'one of the finest soaring sites in southern England'. Lanyon gazed anew on the slender shape of Cornwall. 'The whole purpose of gliding,' he explained in a lecture a few years later, 'was to get a more complete knowledge of the landscape.'

His canvasses filled with flying. Paintings like *Thermal*, *Airscape*, *Cloud Base* and *Silent Coast* were sky-abstracted pictures, infused with a new energy and crispness of colour. In 1960, Lawrence Alloway saw Lanyon's recent work at Gimpel Fils and was amazed by the sense of movement. 'The old works,' he wrote, 'are like puddings compared to the dominant moist and swinging blue.'

But aesthetic fashion was shifting against Peter Lanyon. His themes – the importance of place and tradition, ancient seams hidden in the land – had little relevance in a world of pop art and consumerism. He wasn't swayed. In January 1964, he restated his convictions, reinforcing their seriousness, his attempt not to characterise the undercurrents of the era but to present something ageless and universal: 'I believe that landscape, the outside world of things and events larger than ourselves, is the proper place to find our deepest meanings.' His paintings focused less on the representation of what he saw in his roaming or his flights than on what he experienced. His claim for their appeal was as a version of the sublime. The works came from 'those places where our trial is with forces greater than ourselves, where skill and training and courage combine to make us transcend our ordinary lives'. He defended his continued focus on the land and on Cornwall. It was from marginal sites like those he knew in West Penwith that his calling came, and the excitement they gave him was constantly refreshed:

> Landscape painting is not a provincial activity as it is thought by many in the US but a true ambition like the mountaineer who cannot see a mountain without wishing to climb it or a glider pilot who cannot see the clouds without feeling the lift inside them.

In May 1964, Lanyon took a group of students to the Somerset resort of Clevedon. He had a camera with him and the photographs from those few days – of a place less well-known to him than Cornwall – are stark and abstracted. He produced three paintings which, in their exuberance, their erotic suggestions and their use of colour, suggest a new direction. They have a cleanness about them, too, the neatness of a seaside resort. They all have *Clevedon* in the title.

In August, he returned to Somerset, to fly near Honiton. After one flight, approaching the airfield, he dipped his wing and it brushed the ground. He was slung from the cockpit and broke a vertebra in his back. For a while, his recovery went well; he complained of boredom from his hospital bed – but then a blood clot loosened in his leg, travelled up to his brain and killed him. He was forty-six.

On the final page of one of Andrew Lanyon's scrapbooks is a single photograph. It was taken in the early 1960s by his mother, Sheila. It has that slightly washed-out tone of early colour pictures, the palette of nostalgia for anyone who was around at that time. There is no caption, just a wide-open Cornish beach and a line of icing-white surf in the distance. Peter Lanyon is in the middle of the sand. He is leaning to his right, balanced on one leg. His hands are in his pockets, but they're stretched out wide, so that his anorak spreads open. He appears to be flying. In the left corner – just appearing – one of his children hurries towards him.

From Gunwalloe, I walked north to Porthleven. The path wound along the cliff for miles above Loe Beach, one of the oddest stretches of the Cornish coast. The shingle here is made up of migrant pebbles from all along the English Channel, rolled down distant river valleys in the great ages of glaciation, then shunted up here by the surf, by those 'generations of waves'. At the far end of the beach, the shingle has dammed the river Cober and produced a strange contrast. On one side is a quiet lake fringed with pine trees; on the other the seas roll in from Mount's Bay and flop on to the beach, dragging the water back over the pebbles with a raggedy sound.

I found Andrew Lanyon in his garden above Porthleven harbour. We went inside and sat around a large pine table with his partner, Jacqie. His mother, Sheila, Peter's widow, was propped up in a velour armchair beneath the window, gazing out at something she couldn't quite see.

'What you have to remember about my father,' said Andrew, 'was how Cornish he felt. If he saw someone trying to paint the view out along the coast, he would stride up to them and say, "You do realise, don't you, that there are miners working inside those cliffs."'

Andrew chuckled. He was in his sixties, but had the boyish curls and open face of a thirty-year-old. Ideas and images bubbled from him. He had a striking tendency both in speech and in his own writing to invert perceptions. He mentioned a scene in a short film he was making which involved two fairies who make a rare sighting of a human being. The human is tiny. One fairy hisses to the other, 'But I thought they were supposed to be big?'

Andrew's books about his father are pieces of art in themselves, printed letterpress on mould-made paper. Each image is tipped in by hand. Much of Andrew's own work, including the series about the fictitious Rowley family, is published in the same painstaking way. In one book, a small feather is taped to the page, in each one of an edition of four hundred copies.

Generations of Lanyons have been shaped by the place in which they live – first by its minerals, then by its landscape and its past. It was as if, on reaching adulthood, they each picked up their chair and moved it round to look at the same thing from a different place. Peter joined the artist-incomers, before turning against them. Andrew has produced a series of satirical stories about the artists' colony – in fourteen books (so far). The central character, Walter Rowley, is a scientist who identified art as 'an infectious disease visually transmitted... a vivid fungus which only attacked mankind'. Throughout history, Rowley realises, 'art has repeatedly reached plague proportions, causing civilisation after civilisation to collapse'. Arriving in St Ives, he discovers an epidemic: 'there were so many artists working there that it became increasingly difficult to take a photograph without another photographer or a painter or a sculptor getting in the picture'.

One of his Rowley stories is based on the real encounter that helped launch the St Ives colony. Ben Nicholson and Christopher Wood were visiting St Ives in 1928 when they spotted an old man sitting outside his cottage, painting. Alfred Wallis was a retired fisherman, a widower who had long dealt with his loneliness by daubing bits of timber with pictures. In Wallis, Nicholson and Wood saw the work of a naif, a 'natural' free of artistic tradition – and in that moment the St Ives art colony was conceived. Andrew Lanyon has captioned his picture of the scene: 'Like Montezuma observing the approach of the conquistadors, Alfred Wallis watched Nicholson and Wood, both armed with pencils.'

From her armchair Sheila Lanyon suddenly started talking. 'He wanted to buy one of Alfred Wallis's pictures, you know. Before the war. "What's that mast doing inland?" Peter asked him, and Wallis said: "Mind your own business."' She laughed weakly, and the laugh stumbled to a cough. '"Mind your own business," oh, that'd keep all those critics quiet – "Mind your own wretched business"...'

She fell silent for a moment before continuing, '"If you want that picture," Wallis said to Peter, "you'll have to do one thing. You must read the Bible every day."' A little laugh again. 'He probably did, you know, when he was away at the war, he probably did. Because we had that picture of Wallis's.'

Andrew prompted more, but the clouds had closed in again.

A little later, I rose to leave. 'Bye-bye,' croaked Sheila, trying to find my shape in the fog of her sight. 'I don't know who on earth you are, but thank you for coming to see us.'

Down at the harbour, I walked out towards the open sea and the coast path. The Bickford-Smith clock tower stood tall above the lines of harbour walls and breakwaters. Swells were wheeling in through the narrow gap. In the inner harbour, the day-boats were strung like beads on their fore-and-aft moorings. Children were taking turns to jump into the water, noses pinched and neoprene legs pedalling as they dropped until – *splash!*

In 1950, Peter Lanyon accepted a commission from the Festival of Britain; the subject he chose was Porthleven Harbour and he painted an exuberant jumble of shapes in greens and blues, all topped by the clock tower. The format was vertical – derived, it is said, from the painting he had bought from Alfred Wallis. It makes the harbour look even more precarious, shunted upright by the force of the ocean.

In the body of his work, *Porthleven Harbour* is a transition picture, a stage in the progress from the abstractions that he adopted from the St Ives artists towards his own mission to paint the experience of place. Lanyon wrestled with the painting for a whole year. During that time, he had a row with Ben Nicholson over the 'irreconcilable conflict', according to Lanyon, of trying to create 'an abstract style appropriate to the international scene with a local source'. The Cornwall 'he knew in his bones' could not be abstracted. It was the same battle he'd been having all his working life. Through those months, he worked and re-worked the *Porthleven* canvas, scraping it back so often that it collapsed. In the end he completed it in one burst, in just four hours.

It was early evening. The path ran north and west along the cliffs from Porthleven, a narrow score-line cut into the steep and treeless slope. Far below, the sea hissed around the rocks. The weather was about to break and out to the west, as I walked, I could see grey folds of cloud building on the horizon. The waters of Mount's Bay had that particular agitation that comes before a storm. Ahead of me was Penwith, the last piece of land before the ocean. Above it the sky hung dark and heavy, as over an island.

# PART III

# 14 | MORRAB

From hypothetical Cornish *morrep,* 'seashore'. Borlase notes the division of west Cornwall's parishes into *morrep* and *goonran,* 'upland'.

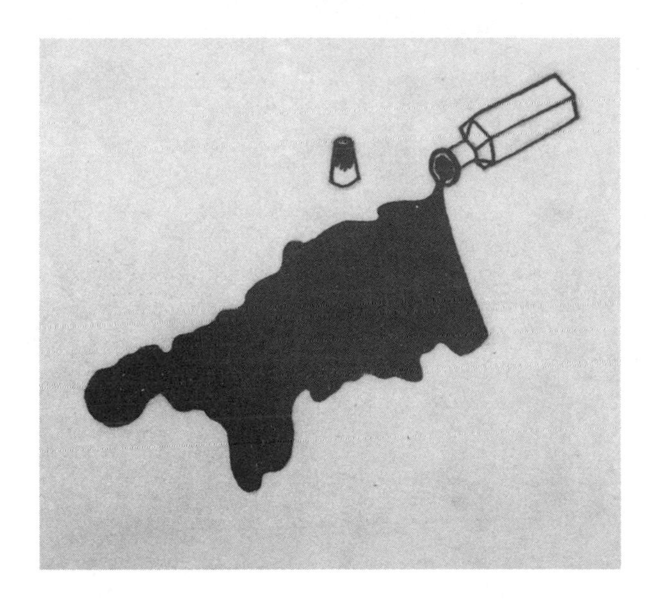

OUTSIDE THE STAR INN IN St Erth is a signboard featuring a map of west Cornwall as it might have looked in the Pliocene era. The sea level was some twenty metres higher then and the last bulge of Cornwall, the old hundred of Penwith, is separated from the mainland not by a narrow strait but by several miles of water. Stylised waves emphasise its out-at-sea isolation and, while I casually read the text beside it – about the fossils found in the clay pits above the village, and how the clay had been favoured not only by miners but by the St Ives potter Bernard Leach – I found my gaze constantly drawn back to that image. There was something deeply unsettling about it: the age-old fear of inundation.

The Penwith peninsula is to Cornwall what Cornwall is to the rest of England – a loosely connected appendage stuffed with the residue of a thousand stories and mythical projections. Every rock, every hill and cliff has its tales, lore and sprites. The peninsula has a mood all its own. 'It is not just a place, it is a *mysterious* place,' said Denys Val Baker, who in 1948 set up the *Cornish Review*. In its first issue, Peter Lanyon wrote of Penwith's extremes: where 'the great and small in life and death [find] animal joy and terror'. Katherine Mansfield was likewise struck by Penwith's uneasy atmosphere: 'It's not really a nice place. It is so full of huge stones . . .' The poet John Heath-Stubbs was more direct: 'This is a hideous and wicked country, / Sloping to hateful sunsets and the end of time.'

Granite cliffs, granite tors, megaliths, the constant flux of wind and cloud and breaking seas – all merging with such force and in such a small area that few people spend time in West Penwith without being affected. But its peculiar presence comes less from the topography than from a cosmological sense of its location. Even before maps, the shape of Penwith, and of Cornwall itself, was obvious to anyone who spent more than a few days sailing the coast or tramping the uplands. The 'corn-' of Cornwall comes from the Cornish *kern*, 'horn' (the Cornish name for Cornwall, 'Kernow', has the same root). 'Penwith' is from *penn-wedh*, in which both elements mean 'end', so it is literally 'end-end', while an early Welsh version of Land's End is *Penpenwith*: 'end of the end-end'. In AD 997 a reference to Land's End appears as *Penwithsteort*, where *steort* means 'tail'. So Land's End is the tail of the end-end, at the end of the horn. You can't get more final than that.

Archaeologists have long noticed something strange about West Penwith. It has the largest concentration of standing stones in Britain. The digging of pipelines and laying of cables is halted much more frequently than elsewhere by the discovery of flint scatters or cists. Few places in Europe can match the number of its hut circles, its barrows, fogous and cliff castles. The funerary monuments are not only numerous but so unusual in form that, in one case, they have earned their own name: 'Penwith chambered tomb'.

It is a powerful coincidence – the end of the land and the proliferation of sacred monuments. No serious theories have attempted to explain it. But heading west through Cornwall on foot is like walking the plank, a feeling made more acute by the mounting realisation that, as the sea approaches, you are also nearing some ritual arena, a testing-ground for the great mysteries, an antechamber to a place that remains always just out of reach.

I had a painter friend who came to spend a winter in Penwith. The thought of it had been niggling at her for years, the necessity of going to the far west. She likened it to having a bit of grit in the toe of your sock, and reaching right down to the end of the sock to take out the grit. To start with, she found herself working with a lightness she had never known before, but then one evening on the cliff beyond Lamorna, she watched low clouds drift in over the sea and felt that each one was smothering her, wrapping her up like a shroud. She was on the train back east the next day.

John Davidson was a poet, dubbed the 'first of the moderns', admired by Wallace Stevens and T. S. Eliot. In 1908, ill and depressed, he fled for the west from London. Stage by stage he moved through Dorset, Devon and Cornwall, until in West Penwith he could go no further. At Penzance he walked slowly into the sea below the town and let the waves close over his head. It was several days before his body was found. 'I felt the time had come to find a grave,' he'd written a few years earlier, 'I knew it in my heart my days were done. / I took my staff in hand; I took the road, / And wandered out to seek my last abode.'

I carried on down to the bridge over the Hayle river, and the crossing into West Penwith. I half-expected a bilingual sign: 'Welcome to Penwith, *Agas Dynnergh Penwith*.' But there was just a poster for 'Salsa Classes' and, on the church hall, a banner for the Alpha Course: 'Life is worth exploring'.

I lingered on the bridge, watching the slow waving of riverweed in the shallows. It all looked as benign as a Hampshire chalk stream. But there are times when a gale drives the spring tide over the mudflats a few miles to the north, on into the narrow gap where the stream comes in; then up here the riverweed stops its waving and flops back the other way. The tidal surge has been known to reach up into the soffit of the bridge. And in those moments, the sea appears to be separating Penwith from the mainland again, prising it away from what is solid.

Walking westwards over the past months, I had managed to follow a vague and not entirely consistent chronology – from the Neolithic monuments of Bodmin Moor, through the Middle Ages and the Enlightenment up to the twentieth-century struggle between modernism and tradition. The timeline ended with Peter Lanyon's glider accident in Somerset. Here in West Penwith all the ages are rolled into one, a post-modernist bundle of residual beliefs, re-interpreted customs, hazy site-myths, ancient stones, recollections and folk tales. One of the best places to begin unpicking it is not out on the moors of the north or along the surf-battered coastline, but in the middle of a municipal garden in Penzance.

Up Market Jew Street, down by the Co-op, into the back alleys behind Chapel Street, past a day centre, and here, amidst the foliage of bushy rhododendrons, chusan palms and tree ferns, is the white stucco facade of the Morrab Library. A bookish fantasy, a private subscription library whose members all seem to be elsewhere, the Morrab is a two-storey warren of high-ceilinged Georgian rooms, with nothing but a couple of empty tables and chairs in each one – and bookshelves, floor-to-ceiling, and high windows which look out over the subtropical garden, over the treetops to the white-crested waves of Mount's Bay.

I spent the afternoon there with a stack of titles rising on the desk. *The Romance of the Stones, The Stones of Land's End, Megalithomania, A Week at the Land's End, The Dust of Heroes.* I thumbed through card indexes, jotted down references and followed more threads than I could possibly tie up. That evening, I took a room in the Con Amore guest house, ten minutes walk from the Morrab, and for several days alternated mornings in the library with forays into Penwith. Late on the first afternoon, I went to see my friend Jeremy Le Grice, a painter who lives a few miles inland from Penzance.

Jeremy is one of my favourite people. His enthusiasm for every little thing that enters his orbit gives him an air of perpetual innocence. Now in his seventies, he had recently been unwell. On the way to his house, I stopped in at Jelberts in Newlyn to get him a tub of his favourite homemade ice cream.

In his late teens Jeremy had studied under Peter Lanyon – a summer of informal classes in a loft in St Ives. These classes left as deep an impression on him as all his years at art school – in part because of the three 'wonderfully sensuous girls' studying with him, and in part because he was swept up by Lanyon's commitment to landscape for what it was, not to its artificial and abstracted form. Jeremy also liked his teacher's irreverent attitude to 'art'. Lanyon told them that if he wasn't pleased with a canvas, he'd lay it on the ground and drive over it in his car. 'Usually improved it,' he'd say.

Jeremy shared with Lanyon a long family connection with west Cornwall. He had spent many years away, but for the past two decades or more had worked from a studio in Newlyn, painting some of his best work. Here in his panelled sitting room hung one of his nocturnes, and I remember him telling me how, for a whole winter, he had spent the evenings looking out of his studio window at the shadowy shapes of Mount's Bay. Ever since, I've always pictured him like that – a lone figure in the yellow glow of his window, high above the sea.

Now the warm June sun flooded his sitting room. He sat with a moss-coloured rug on his knees, the fire crackling in the grate beside him. He opened the tub of ice cream. With a tired smile, he scooped the first lump into his mouth. He had slowed since I saw him last, but when we talked of Penwith he was as expressive as ever.

'There *is* something about this place. The more I think about it, the less I can really put my finger on it. People talk about the light – "Oh, the light of St Ives," they say. But that's nonsense. It's something innate. I think of it as a secret buried in the land. It's not something you see – it's something you *feel*.'

We spoke of mutual friends, and of the persistent myth of the Phoenicians in Penwith, and of our own shared ancestor – a naval man from the nineteenth century, known in both our families simply as 'the admiral'. He had had a dashing early career sailing gunboats up the Amazon, into China, in and out of Italian ports in support of Garibaldi. Late in life, he married for the second time – to a very young Italian from the hills above Ancona. Jeremy's eyes sparkled at the memory of his visit there: 'Did I ever tell you about going to their village? We found one of his descendants, amazing woman. They put on a huge feast for us! A huge feast...'

He finished the ice cream and carefully placed the empty tub on the table beside him. He looked down at the fire. The flames flickered on his cheeks, but there was a blankness in his face that I hadn't seen before. 'I haven't really been painting recently. I've become rather passive, Philip. But a couple of weeks now, and I should be over the worst of this trouble.'

His wife Lyn showed me out. We walked through their walled garden. There was a scent of jasmine and lavender, and overhead the chatter of rooks in the treetops.

'I'm afraid he won't be over it. It's here, everywhere – ' she put a hand on her middle. 'They can't operate.'

Jeremy died just a fortnight later.

In the basement of the Morrab Library was a small and windowless room where Mr Simmonds was cataloguing and preserving the library's archive. He was a compact man, which was just as well, as there was little space to move down there. High shelves filled the room, holding framed prints, sketchbooks and canvases, bound volumes of local newspapers, parish records, acid-free boxes of letters, manuscripts, personal papers; and, in several stout cardboard containers, beneath layers of tissue paper, the letter books, the ledgers and manuscripts of Cornwall's greatest contributor to antiquarian studies.

Born in Penwith in 1696, William Borlase was a man of his time: a polymath whose passion for the world drove him to a meticulous recording and analysis of everything. He did more to commit Cornwall to paper – its minerals, creatures and its plants, its landscape and its antiquities – than anyone before or since. He was also a member of that group – the ancestral spirits of modern archaeology – who changed for ever the way we view the land and its past.

His letters reveal the astonishing range of his interests. He studied the 'Physick'. He became 'fond of Chemistry'. A copy of Linnaeus's *System of Botany* set his 'heart a'gadding after the natural curiosities'. He took up drawing. He corresponded with Alexander Pope, who sent him a copy of his work in exchange for a batch of glittering Cornish rocks (mineralogy was another enthusiasm). One thing Borlase did not pursue was poetry, having had a disastrous dalliance with versing as a student, when he 'writ a couplet or two on my Oxford mistress's stays; I was soon after discarded by the fair one and laughed into my senses by my companions'.

After Oxford, William Borlase returned to his native Penwith. He took up the living of Ludgvan, married a suitable woman, had half-a-dozen children and settled into a round of such contentment that, reading of his life then, it's a wonder he ever picked up a pen for anything more than a platitudinous sermon. At the age of thirty-one, he concluded:

> I have had the pleasure of seeing some of the most considerable places in England, and I think there is hardly any place I could so willingly wish that my lot had fallen as where it has . . . The gentry, most of whom are our near relations, are of a free, frolicking disposition. In the Summer time we meet (some ten or a dozen) at a Bowling Green, there we have built a little pleasure house and there we dine, after dinner at bowls, and by so frequently meeting

> together we are as it were like so many brothers of one family, so united, and so glad to see one the other.

Another two decades passed. Still with nothing published to his name, Borlase noticed his energies starting to dim. He understood, without regret, that the chance to make anything more of his life had already passed:

> I am preparing for old age, that is, laying in a fund of amusements, such as may enable me to spend my time within doors to my satisfaction…I read a little, I write a little, I paint a little, I collect a little, I think a little unless it be upon my friends, and them I hope I shall never forget.

But one day in May 1748, he happened to bump into two distinguished antiquarians in Exeter – Rev. Dr Charles Lyttelton and Rev. Dr Jeremiah Milles. As he spoke to them of west Cornwall's antiquities, the sites he knew from his own amateur inspections, Borlase watched their amazement. Letters followed the meeting, and Borlase wrote back with details of the places he'd studied, and soon the word spread and others wrote to him, flooding the rectory at Ludgvan with letters and connecting him with that circuit of scholars from all over Europe who were opening up the earth's mysteries. Borlase became convinced not only that Cornwall and West Penwith in particular were places of significance, but that it was his duty to reveal them.

So at the age of fifty-two, William Borlase began his life's work. He wrote to his fellow parsons around Cornwall: 'Have you any British or Druid Antiquities; any rude obelisks of stone, either single, or in a straight, or circular line? Any seals, amulets, rings, or bracelets? Any cromlechs? Any basins cut into the surface of your rock?'

Response was limited; he ended up doing most of the research himself. He covered hundreds of miles, chasing up and recording every trace of antiquity. Many of the monuments have since been lost, and his diligent surveys often remain the only record of their existence. He complained of being too far from libraries, but his essays are filled with Biblical and classical references, along with quotes in Latin and Greek and Hebrew, Gaelic, French and German, further expanded with gobbets from British topographies, Ancient Egyptian histories, Persian cosmologies, Scandinavian ethnographies, and citations from esoteric literature from Paracelsus to John Dee.

Borlase's *Antiquities of Cornwall* is vast, as revealing of its own age as the structures described are of theirs. The Morrab has a facsimile of the second edition, which was revised by Borlase himself. With both hands, I heaved it up the open staircase – beneath the high gaze of John Borlase of Pendeen, William's father – to my chosen desk in room no. 8. After a morning spent squeezed into the basement with Mr Simmonds, here was sunlight falling through the high sash window, sparkling on the waters of Mount's Bay and honeying the paragraphs and engravings of Borlase's work. His *Antiquities* was soon joined on the desk by his equally compendious *A Natural History of Cornwall*.

When William Borlase looked at the world around him, what did he see? He saw what we see. He saw the same moors and cliffs, the same high tors, the same field boundaries, the same woods (more or less). He saw the same shifting weather (and recorded it for decades, twice a day, in tabular form). He saw the same birds (for fourteen years he kept a pet chough). He saw the same rocks, the same crystals and minerals from the mines (his study of Cornish 'fossils' earned him election as a Fellow of the Royal Society). In the harbours, he saw in the bilges of the punts and mackerel drivers, and, in the rows of strand-line maunds, the same fish, which he spent hours drawing – blackfish, sunfish, the weaver or 'sea-dragon', the pipe-fish or 'sea adder'. He saw the same standing stones and stone circles, the same strangeness and diversity, and, as he travelled around the lanes and over the moors of west Cornwall, he felt the same curiosity and wonder.

But the past through which he understood it all was different. It was a lot shorter, for a start: the earth began just 4,004 years before the birth of Christ. All living creatures were descended from those that survived the flood (in the archive here in the Morrab is a paper Borlase wrote, headed 'Private thoughts of the Creation and the Deluge'). The people of Europe were the offspring of Noah's son Japheth. Borlase's minerals were all called 'fossils'. Even six centuries after Geoffrey of Monmouth, the British History is not quite dead: 'It could not be wonder'd at,' Borlase wrote, 'if some British customs were like those recorded of the ancient Trojans.'

And in the ancient ritual sites, Borlase saw the hand of Druids. Everyone did. The Druids had been a part of prehistory since the rediscovery of classical histories during the Renaissance. Accounts of Druidic practice and belief crop up in the work of Posidonius, Pliny and Caesar, who regarded them as little more than a savage and recalcitrant priesthood. By the mid-eighteenth century, though, 'the mood was shifting', wrote the archaeologist Stuart Piggott, 'from rational to romantic, from classical calm to barbarian excitement'. For many, the Druids became something of a cult, and for none more so than William Stukeley.

Slightly older than William Borlase, Stukeley's Druidic interpretation of British antiquities dominated the field then – as for some it still does. Taking the title of 'Chyndonax, Arch-Druid', Stukeley travelled around Britain studying ancient monuments. He was as tireless in his pursuit of the past as John Leland, and was driven by the same heady patriotism, the conviction that Britain's secret glory lay in its antiquities. For six summers Stukeley surveyed Avebury and Stonehenge. One evening, he dined on top of one of Stonehenge's lintels. Avebury, he declared, was 'the most extraordinary work in the world'. It was also deliberately built in the form of a circle and a snake to represent the Father and the Son. Stukeley became convinced that such sites proved Britain was the seat of the patriarchal faith, the ancient belief of Abraham, and that the Druids were the keepers of this lost and primal wisdom. It was Britain's destiny now to reveal it again to the world, and create the New Jerusalem. Stukeley's work fired William Blake's millenarian vision:

> When shall Jerusalem return & overspread all the Nations?
> Return, return to Lambeth's Vale, O building of human souls!
> Thence stony Druid temples overspread the Island white . . .
> All things begin & end in Albion's ancient Druid rocky shore.

The Druidic idea was the direct successor to Geoffrey of Monmouth's British History.

Stukeley never reached Cornwall, but he heard tell of its wealth of monuments. He realised that their abundance in the far west must indicate the place where Druids first arrived in these islands. The man to consult was William Borlase, and their correspondence survives in the Morrab archive. I spent the following day down in the basement, juggling the boxes in that cramped crypt.

Borlase's letter books of the time lay in a box at the bottom of a pile. On 17 October 1749, Stukeley asked Borlase to measure the Hurlers and Boscawen-Un stone circles and the nearby Men-an-Tol. 'I am thoroughly persuaded our Druids were of the patriarchal religion, and came from Abraham,' he wrote to Borlase. 'I believe Abraham's grandson Apher helped to plant this island and gave name to it . . . I am exceedingly indebted to you,' he gushed, 'and especially since you are a co-operator with me in the same argument.'

But Borlase wasn't. He found the Druids faintly repellent, with their pagan rites and human sacrifice. In his reply he added a hint of scepticism about Stukeley's 'conjecture'. Stukeley took offence and there were no more letters. But in their brief exchange, the great chasm between the two men was clear. Borlase was modest about his work: 'I thought some things might be added to the accounts I met with, from a faithful measurement, and observation.' Stukeley on the other hand brought his own theories to the evidence. He drew a picture of a preconceived arrangement of stones – a 'kobla' – and asked whether the remains at Boscawen-Un might be made to match it. He queried the accuracy of Borlase's measurements because they did not correspond to the 'Hebrew cubit'.

I put away Borlase's letter book. I placed it back on top of a similar volume in its specially made, cloth-covered box. Locking the door of the archive, I climbed up to the ground floor and out into the Morrab Gardens. The sun was bright and the south wind felt firm against my face. Fresh air! I walked through the gardens down to Chapel Street, where I had an appointment at the Admiral Benbow with a group of Penwith's pagans.

# 15 | MADRON

From Saint Madern – Madron is still sometimes pronounced 'maddern'. Nothing is known of the saint's life, but there may be a link to the Welsh Madrun, also linked to Saint Materiana, patron saint of Tintagel church. An alternative name for the parish is Landythy: *Lann-*, sacred enclosure; *-dythy*, similar to the *-ith* of St Teath, a corn goddess.

The admiral benbow lived up to its Treasure-Island name. It had a squat facade and clandestine windows. Along the roof crawled a ruffian sailor in painted plaster-of-Paris. Three men were standing on the pavement below, smoking, as they discussed the Spanish invasion of Mousehole in 1595. Each of them was wearing a bowler hat.

'Tuesday's bowler-hat night.' Inside, the landlady pulled me a pint of Betty Stoggs. She herself was dressed in an open-backed dress of black chiffon.

'I thought it was pagan night?'

'The pagans too.' She handed me my change. 'Upstairs, love. You'll find them in the back.'

At the top of the stairs was an unlit room. I stumbled into a pile of stacked-up chairs, part-spilt my drink and felt my way through several unused rooms to an open area at the back, with two walls of windows.

The pagans were gathering, arranging themselves on the chocolate-brown banquette seats. They were a diverse group in age and dress, and we nodded and smiled at each other in the way of waiting for something to begin. In the middle of the room the group's leader, Sarah, was lighting a ring of tea-lights on a table, which was covered in a magician's moon-and-star cloth. There were nineteen candles – the number, she explained, of the stones in West Penwith's stone circles.

The room itself was painted in fading fairground colours. The columns were barley-twists of gold, topped by swags of fruit in red and green and blue. Fairy lights ran along the beams overhead; a projector was mounted on the ceiling. It had the musty smell of long-ago parties. On one wall was fixed the bow-section of a ship's lifeboat. I stood and went closer to read the name, cargo-stencilled beneath the gunwale. The *Torrey Canyon*! That was what the summer of 1967 meant here – the tanker with its back broken on the Seven Stars Reef, leaking crude into the Western Approaches. A flood of memories washed over me, beach memories – of tiny oil-globs on flip-flops and towelling shirts, and black sticky patches on the rocks at the top of the tide.

Sarah called the moot to start. We formed a circle around the candles. We held hands. We closed our eyes. 'Breathe out—' A woman's voice settled over us, strong and resonant. 'Take three breaths . . . One for the sea that surrounds us . . . one for the sky above us . . . and one for mother earth that supports us . . .'

Sarah read out the notices. There was to be a regional gathering of pagans in Bude. Ronald Hutton was speaking. There was a chorus of excitement about the thought of Ronald Hutton speaking. Who could put up posters? And their own festival, down here in West Penwith – might it be inside, with the weather being what it is?

'What!' joked one woman. 'Are we fair-weather pagans?'

A man with an impressive forked beard stood and announced a clean-up at the ancient round-house settlement at Bodrifty. 'Be a bit muddy along the path, mind – but 'tis a lovely spot.'

Another woman said she and a friend had been up at the holed stone of Men-an-Tol, clearing the drainage trenches. 'Coachload of Germans turned up. What must they have thought of these two mad-women coming up out of the ditch, waving and saying hello – all covered in rab!'

Beside her a quiet-spoken woman said she'd been clearing up the offerings at one of the holy wells. 'When will these people learn to leave something biodegradable?'

The day faded outside. The moon rose pale over the land,

spreading its glitter across the sea. The castle-topped profile of St Michael's Mount stood black against its glow. Some of the group went off to get another drink. When they returned, the talk turned to mandalas and their links with stone circles. They asked me of my interest and I spoke of the power of place and how it had altered over the centuries, and how it stayed the same. 'Perhaps for the first time in five thousand years, we are seeing the Neolithic monuments in the way they were intended.'

That prompted a youngish woman to talk of a group of mini-quoits in an area of mine-spoil near Pendeen. Each time she went, there were more. 'They're beautifully made – and no one seems to have a clue who's doing it.'

Another man, a builder, told a story of laying a maze for a client and going back after he'd finished and being so affected by the site that he had to sit down against a tree. 'It was the owner who found me there, having a kip!' A woman spoke of a cairn-field on the Isles of Scilly and the piles of stones left by visitors. 'Wish they wouldn't do that.'

What was common here, I thought, was not so much belief or rite – 'put five pagans together,' said Sarah, 'and you have seven different versions of worship' – but a deep affection for the sites themselves.

There was a man in the shadows. He'd been silent until now. When he spoke it was softly. He said he was a Cornish speaker, from Redruth. His voice had a lovely song-like lilt. 'Granfer took me up Carn Brea one afternoon, years ago now, musta been eight or so at the time. He kicks back the grass round the top there and grabs my 'ands and presses them down into the bare soil. "Feel that, boy? Does 'ee feel it?" I felt nothing but the mud. "That, boy! 'Tes the beating heart of Cornwall!"

'Well, I didn't dare move – with my hands down in the peat there and granfer pressin' them down and then he lets go and leans in close and says, "But you don't go telling no one, boy. Specially not Grandma – she thinks I'm a good Methodist!"'

Back in the Morrab Library the next morning, I found a small booklet which, in its opening lines, nailed its colours to the mast:

> Cornwall! The Delectable Duchy! The Fairies Playground! The Land of Junket and Cream! The Land of the Giants! The Land of the Saints! To these titles we would add one more – CORNWALL: THE LAND OF THE GODS.

The booklet was written by T. G. F. Dexter, a Cornish antiquarian who published a number of works of old lore between the wars and was convinced that Cornwall remained more polytheistic, more pagan, than anywhere else in the country. His style falls somewhere between the pomp of James Frazer in *The Golden Bough* and the free-ranging prose of Robert Graves in *The White Goddess*. A good deal of it is pretty far-fetched – lots of speculative talk of water-gods and fire deities and horse cults. Some of the toponymy is a little dodgy. But its sixty pages and one hundred and thirty numbered points add substance to one of Cornwall's abiding impressions – of elemental forces so powerful that even the most agnostic of souls can be set a-quiver at certain sites.

Two-thirds of all west Cornwall's parishes, says Dexter, had fairs that coincided with days of traditional sun worship. He quotes the folklorist M. A. Courtney, who noticed that during Holy Week 'every vehicle was engaged to take parties to some favourite place', invariably the sacred sites of old. And many of those Cornish saints who live on in the names of parishes were not saints at all but derived from pre-Christian cults. St Michael Penkevil is St Michael of the 'horse's head'. St Michael himself often equates with the sun deity, and an old name for St Michael's Mount is *Din-sul,* or 'hill of the sun'. St Allen was not originally a man but *avallon,* 'apple-tree'; St Teath comes from *ith,* the corn-goddess.

Oliver Padel is the current authority on Cornish place names. Checking his toponymic dictionary threw up more site-derived 'saints': St Dennis in the clay country comes from *dinas,* 'enclosed fort'; and the island of St Agnes on Scilly is from *enys,* 'island'. Charles Henderson, noted Dexter, had spotted the same tendency: 'one of the greatest of Cornish scholars admits the possibility that saints' names may have been sometimes compiled from place names'.

Site-based belief can be as tenacious as a limpet. Proselytising faiths like Christianity and Islam depend on claims of universality and from time to time have had to prise their followers away from the subversive reverence for particular places. During the Reformation, zealous Protestants set out to purge Britain of its pagan sites. Holy wells were blocked. Way-chapels were destroyed. Stone crosses were smashed. Pilgrim rites were banned. Monasteries – with their pilgrim-bait of holy places, relics and cures – were stripped. A century later Cromwell's New Model Army carried on the good work, burning the winter-flowering hawthorn at Glastonbury – the 'Christmas thorn' – and even banning Christmas itself for its pagan associations. If scriptural backing was needed, the enthusiasts pointed to the fourth chapter of St John's gospel in which, at Jacob's Well in Samaria, a woman asks Jesus about finding the sacred in the land: 'Our fathers worshipped in this mountain,' she says. And Jesus replies: 'God is a spirit: and they that worship him must worship *him* in spirit and in truth.' Calvin taught that God had 'abolished all distinction of places'. Those who made journeys for pious reasons were no more than 'spiritual fornicators'. The Biblical translator George Joye spoke for many Protestants when he admonished those 'who runne after straunge goddes, into hilles, wodes and solitary places, there to worship stickes and stones (yea & peradventure to do worse things)'. Elevating certain places for worship was seen as no less idolatrous than kneeling before a golden calf.

It is the age-old argument. Some consider the glories of the material world as the clearest conduit for religious expression; others see them as the greatest impediment. For most of the time the two urges bumble along in that state of latent hypocrisy that is a requirement for any established faith. Zoroastrianism is usually seen as the fount of the dualist tradition that identified the physical world as the work of the devil – yet visiting the six *pirs*, the sacred sites around the city of Yazd, is an important rite. Muslims resist the idea of incarnation, abhor the image, and repeat the Koranic mantra 'There is no god but God' – yet one of the five pillars of Islam is to reach the sacred site of the Kaaba in Mecca, itself a representation of three hundred and sixty pre-Islamic deities.

I once spent a week or so in the non-state of Abkhazia on the eastern shores of the Black Sea. It was the late-1990s and the secessionist war had left it in a shell-scarred limbo between Georgia and Russia, but what drew me there was its extraordinary botanical and ethnic resilience. A total of one hundred plants are endemic in Abkhazia; I visited a glade where a botanist bent to show me the fronds of a delicate fern – one of only five survivors of its species. Abkahzians themselves are alleged to be some of the longest-lived of any people in the world. In terms of belief, the entire country is a sacred landscape, centred on seven holy sites. Traditionally, each family also had its own *a'nyxa*, a particular ritual site in the hills or forest where they would gather for prayers and ceremonies. During the war with the Georgians, and leading up to it, visits to the sites increased. (I had noticed something similar during the Karabakh war further south: *khach'kars*, Armenian stone crosses erected in remote spots attracted more offerings, more candle-burning and toasts the closer they were to the fighting.) In Abkhazia, a friend took me to his native village and pointed to his family *a'nyxa* high up the slope. When I asked to go there, he shook his head. 'It is mined.'

One explanation for all three aspects of Abkhazia – its endemism, its longevity and sacred sites – is a stability that has allowed exceptional levels of adaptation, in ecosystems as much as among the people themselves. The climate has been both consistent and favourable since the Palaeolithic. Abkhazians have lived here thousands of years, growing and eating the same food, building resistance to the same diseases. They have never had to pack up their gods and take them away from the land that produced them. Faith is still attached to places. While the monotheisms have set up camp in Abkhazia, they have not driven away the ancient loyalty to sacred sites. It is often said of the Abkhaz people that they are sixty per cent Christian, thirty per cent Muslim and one hundred per cent pagan.

The following day, in the early evening, I left the Morrab and walked north to Madron Well, one of the most persistently revered natural places not just in Cornwall but in the country. Like prominent hills, water sources have been an object of worship all over the world; unlike hills, they also have a practical use. The vigorous flow at Madron met Penzance's growing demand for water in the seventeenth and eighteenth centuries.

The evening was windless and warm, the verges a-buzz with bees. Beyond Madron Churchtown, they were cutting hay and the air was thick with its herbiness. Lines of grass ran around the shorn fields; from a distance it looked like an elaborate maze. Off the main road, there was a sign, Madron Well & Baptistery, and I followed a muddy path beside an area of dense carr, where willows pushed their serpentine boughs in and out of stagnant pools. Sunlight found little patches of marshy ground and glistened on the pools. I came to a drinking place, where the ground was muddy and well-trodden.

But it wasn't cattle who came to the bog here and left the trees hung with strange fruit. Everywhere you looked, the mossy branches were covered. Ribbons and rags and clouties, a crimson hairband, white lace, blue velveteen, a gold and red friendship bracelet, a three-flag shred of scarlet bunting, a medallion with *Your Guardian Angel* written on it. Such an outpouring of hope! Yet there was no one here. It felt like the camp of a persecuted cult, abandoned in a hurry.

Who had tied this here – a shred of cotton in bright autumn paisleys? Or this ladybird button, or this length of petrol-blue wool in a neat bow? Or this laminated label printed: *13th Prize*? And what had become of the lonely soul who'd written on an index-card: *please bring me love*? And the patient prescribed this foil strip of a course of omeprazole (20 mg)? Or the baby whose tiny moccasin hung by its laces, the caramel leather now patinous with mould? And what had happened to poor Lily, a prayer for her recorded in a confident hand on a brown luggage label: *Ballerina Songs / With Butterflies and flowers / for Lily*?

We know what happened to the soldier William Cork, who was cured long ago by drinking from the well and carried on with his campaigning for the Stuart kings; and to John Thomas, fisherman from the same era, rid of his lameness and able once again to go to sea; and to the local parson with his 'cholical pains' who stopped a woman walking back from the well, downed her bottle of Madron water and – despite scoffing at its powers – was relieved of the pains. We know that King Charles I was told of Madron Well and John Trelille, and asked for an investigation. During a game of football, the young John Trelille had run off with the ball and was attacked by a girl with a stick. She whacked him so hard he ended up crippled, spending the next sixteen years walking with his hands, dragging his useless legs behind him. But the waters of Madron Well cured him. He was soon back on his feet and fighting loyally for the King in Dorset, until killed at the siege of Lyme Regis.

Over the years Madron Well's sanctity has, like the Neolithic monuments, produced a whole host of different reactions. William Borlase, ever reasonable, expressed a pragmatic scepticism. To Madron Well, he wrote, 'come the uneasy, impatient, and superstitious [but] instead of allaying, deservedly feed their uneasiness; the supposed responses serving equally to increase the gloom of the melancholy, the suspicions of the jealous, and the passion of the enamoured.' Others have found the well an abhorrence. A detachment of Cromwell's men destroyed much of it during the Civil War.

Beyond the well stand the ruins of the twelfth-century baptistery. A chest-high wall encloses a space little bigger than a garden shed. In one corner of it, water presses from the rock and runs out over the floor. At the other end, a bunch of campion and stitchwort lay wilting on a crude granite altar. Above the altar the top course of stones of the wall had been removed. I looked at it for a moment before realising: the stones had been *ripped out* (a report by a local conservation group later confirmed it).

The sacred places of West Penwith still provoke extreme reactions. A few years ago, wooden stakes were found driven into the ground at the stone circles and beside various standing stones. The granite had been doused in red wine and on the stakes was written: JESUS. One stake also bore the words from verse twenty of the thirty-third chapter of Isaiah, a stern reminder that there was only one place on this earth whose veneration is not idolatrous:

> Look upon Zion, the city of our solemnities: thine eyes shall see Jerusalem a quiet habitation, a tabernacle that shall not be taken down.

Here at the baptistery, the scar left by the removed stones was dry and dusty; white roots dangled from it. The cavities were like a missing row of teeth. What makes someone desecrate such a site? The insecurity of their own faith? The fear that a belief other than their own might take over? In that dank ruin it was clear that it was not believers who were being attacked so much as the site. I found it astonishing that the place where water seeps from the ground can exert such power.

# 16 | ZENNOR

From St Senara, possibly the Breton princess Azenor who gave birth to St Budoc in a barrel, while drifting across the sea from Brittany.

On a grey morning towards the end of that week, I left Penzance to walk the final section to Land's End. Another summer storm was driving in across the Atlantic, blowing hard from the north-west. The wind dashed over the peninsula, to skid out into Mount's Bay in dark, darting patches. I headed inland, and for some hours happily followed the back-lanes, listening to the hiss and pull of the gale in the treetops.

Then there were no trees. To the west of St Ives rises a tract of moorland that runs all along the top of Penwith. Climbing through bramble and scrub, then tight-packed gorse and heather, I came out out on top of Trendrine Hill. Beyond the sea was a mile-high mountain of dark cloud; sunlit against it, the trig point looked like a miniature floodlit church. A few hundred metres to the west was a granite tor, topped by a vast egg-shaped rock. Scrambling up the boulder blocks beneath, I stood before it. In the top of the rock was a bowl-like cavity. I ran my hands around inside and was amazed by its regularity. It looked hand-chiselled.

Far below, stretching out to the horizon, the water was scuffed white, and blurred in places by coming squalls. Half a rainbow dropped from the cloud – less a rainbow than a curved chromatic lance thrust down into the water. The rain came suddenly – hard and stinging. I ducked my head and watched the first hailstone strike the rim of the bowl and bounce around inside it like a ball in a roulette wheel.

When William Borlase studied these curious rock-basins – the first to do so – he was quite clear about their liturgical function: 'From these Basons, the officiating Druid standing on an eminence, sanctified the congregation with a more than ordinarily precious lustration.' The priest would stand before the basin. He might drink the water straight from the rock, or he might wash his hands in it, or rinse oak leaves or sprigs of mistletoe (whose holiness, like the water, derived from its never having touched the bare soil). Sometimes the basins were carved into logans, or 'rocking-stones': the miraculous sloshing of the holy water as the stones moved was enough to 'make the criminal confess; satisfy the credulous, bring forth the gold of the Rich.' There are dozens of these basins on high granite outcrops on the Penwith moors, on Bodmin Moor too. Often they are on the same rock, in groups, linked by tiny runnels where liquid could flow from one pool to another, invariably ending in a larger reservoir.

Borlase began his discussion by disposing of suppositions. The basins were not, as some thought, used by tinners for pounding the ore – why would they be in such high places? Nor were they water-bowls for the dogs of King Arthur. They were the wrong size or shape for post-holes or sockets for obelisks or stone deities. No, concluded Borlase, the basins were used for 'Ablutions, Lustrations and Purifications'. They were designed to collect the purest water for this purpose, in the form of rain, snow or dew. The rock-basins, he was convinced, were carved by men for ritual use.

Unfortunately for Borlase, they were not. They were formed by the slight acidity of rainwater degrading the granite matrix and loosening its feldspar crystals. It was not Druids who made them, but frost action during the periglacial period.

We know that, now. But in prehistory they did not, any more than Borlase did in the eighteenth century. The point is that they look man-made. So Borlase's work is perhaps not so misguided. It is easy to imagine people coming to the basins in the early Bronze Age and reading in their regularity the work of distant ancestors, and 'reusing' them.

One of Borlase's longest sections in his Druidic essay is entitled 'Of the Druid Places of Worship'. His own fieldwork had revealed sites where not just one but a whole series of ritual monuments were clustered. Rock-basins were often found near cairns, which themselves were adjacent to each other: 'It will perhaps seem surprising to many readers, that many places of devotion, and Altars of the same kind, should be found so near one to the other; Karns for instance, on adjoining hills... but it must be remembered, that the ancients were of the opinion, that all places were not at all times equally auspicious.'

As a good monotheist, Borlase enjoyed pointing out that if the Druids – these 'superstitious jugglers' – sometimes failed to make the spirits manifest, they blamed not their own powers but the place itself, and moved to another location. In his work on rock-basins, Borlase identified – both in the field and in classical literature – the principal framework that archaeologists now use to explain these features. Whether such features were naturally formed or man-made matters less than how they were *perceived*. The stone monuments of antiquity can be understood not so much by their design, or by the study of excavated objects, as by their siting, their relationship to other sites and the suggestiveness of their form.

*'All places were not at all times equally auspicious...'*

In the mid-eighteenth century, William Borlase had identified the idea we now know as 'ritual landscape'.

The squall passed quickly; the sky cleared. I carried on along the ridge. Below me, the flat ground along the coast stood out like a street map – grey, treeless walls around little blocks of green, all beneath the blue distance of the sea. When W. G. Hoskins wanted an example of what land use might have looked like in England before the Romans, he came here, to the parish of Zennor. The ancient fields are among the best-preserved in Europe, still used for exactly what they were built for some three thousand years ago, giving the farms a particular aura of antiquity.

In Zennor itself, an old water-wheel tinkled as it turned outside the Wayside Museum. The collection of farm tools inside was largely gathered in the 1930s by one Colonel Frederick Hirst, lately retired from India and having long ago changed his name from 'Frederick Shirt'. The story goes that one day the colonel was waiting for a ferry on the quay at Hayle when he spotted a pile of scrap metal. Going closer, he saw that it was largely made up of agricultural tools headed, he was told, for Germany. Hirst was less alarmed by the thought that they might be converted into armaments than by an entire way of life vanishing with those tools. He arranged to buy them, and sent them back to West Penwith, to his new home near Zennor. It was the start of a project that consumed him for years.

Hirst visited every farm in the stony labyrinth of Zennor's fields. He rummaged through barn corners and lofts, the nettle-forested fringes of abandoned pig-yards. After reclaiming one hoard of tools from a shed, he was told: 'Them's nothing. You can have the lot – except the thristle [flail].' Hirst not only collected utensils; he recorded local particulars too. His archive in the Courtney Library contains an intensely detailed typescript survey of the entire parish – field names, sizes, maps and descriptions. Having had a successful career studying the revenue patterns of the states of Bengal and Bihar and Assam for the India Survey, he now developed a passion for Europe's half-buried past. He hunted antiquities in the Pyrenees and visited the great archaeologist Abbé Breuil. In Cornwall he also started to organise local digs, and his West Cornwall Field Club was the forerunner of the Cornwall Archaeological Society, which is now one of the best and most active such societies in the country.

Wandering through the warren of the Wayside Museum, I marvelled not so much at the tools themselves (there's only so many augurs and ploughs you can take) as at the thousands of years of activity they represented on this remote coastal plateau. Like Charles Henderson, Colonel Hirst was horrified by his age's carelessness of the past, by the eradicating of all that made one place distinct from another. Neither man is a part of the roll-call of national figures, but in their modest way their championing of the local, their snatching of records from the flames, makes them look like heroes now. 'A little thing is a little thing,' said Hirst of his collection, quoting St Augustine, 'but a little thing well done is a very great thing.'

I found a B&B in the hamlet of Trewey, just outside Zennor. Mrs Mann wasn't from 'round here', she said. She gestured away westwards along the road. 'Out Morvah way.' She had married into a Trewey family but just a year ago her husband had died and now the farm was worked 'by our son and his wife, and by his son and his wife.'

We were standing in her sitting room. 'So, you're from up Truro?' She was inspecting me; she had a kindly, open face behind pertly-rimmed glasses. ''Spect it's all Friesians up there.'

Her family kept Guernseys. They were better suited to the small fields. We talked about the herd for a while, and about a coachload of tourists from County Cork who'd driven past last summer and found it so reminiscent of some parts of Ireland that they stopped to have a look. 'Course, we got talking – and, you know what, I'm going over there this summer to visit!'

She then told a lovely story about one of the family's cows. 'A man from the Cattle Society was passing and he spotted the cow and he said to us: That's a beautiful cow there – you should show her. Well, we said, thank you, but we're not show people, not really. But later we thought, why not? We put her in the local show – and she came first! And they said there at the show, you should take her up Royal Cornwall. We told them we're not show people, but when it came to Royal Cornwall time we thought we'd take her anyway. Swept the board, she did! And they all said, you should take her up the nationals. That's up-country – Coventry way. And bless me if she didn't win that, too – champion cow of all England!'

In the kitchen was a photo of her – a rust-red beauty queen, side-on and proud in a show-ring. 'Look! She loved having her photo taken. Loved all the people looking at her.' I peered in close, and saw in her glistening brown eye the faint look of pleasure as she drank it all in.

I was Mrs Mann's only guest and the next morning had breakfast alone in her dining-room. Everywhere were crammed pictures and ornaments. All were rooted in the land in some way – a wedding picture of her grandson: bride and groom standing in a just-mown field. The walls were hung with horse-brasses, embroidered pastoral tableaux, a framed jigsaw of a harvest scene, and two identical prints – one in each of her front rooms – of Elijah being fed by ravens.

Mrs Mann stood on the doorstep and waved goodbye. The fog had come in that morning and when I turned around further up the road, I could only just make her out, and the shape of her hand, still waving.

All morning the fog ebbed and flowed. On the moor above Bosporthennis, I could see no more than fifty metres ahead, now a hundred, now twenty. Each step brought a bog-quiver of peat beneath my boots. The wind soughed in the gorse. The world all around was white and the white was without form, and void, and it did tricksy things to my thinking. I'd been walking west when it thickened, but now with the twists of the path, I wasn't so sure. Rain began to fall.

When the cloud parted, it did so suddenly. Three things appeared at once, like a cast at curtain-up. Half a mile to the south: Ding Dong mine – its long-abandoned engine house and stack a fist and finger pointing to the sky. To the north: the summit of Carn Galver – ritual focus of the area's ancient monuments. And ahead, on a small knoll: the scattered boulders of an opened barrow. Looking around, I spotted beyond it the surviving markers of Boskednan stone circle.

There is always a threshold moment when you step into a stone circle, a sense of entering something, however few the stones. Boskednan has just eight. You glance at each of them. You count them. You fill in the gaps where they are missing. You might go up and touch them, feeling the stone cool against your palm. You might sit down, prop your back against one. Or you might busy yourself with pacing and dowsing or with spotting alignments. You might simply amble about inside. But whatever you do, and however sceptical you are about their mystique, you will find questions slipping into the silence: who, how, when – and *why*?

Barrows and quoits and stone rows all suggest purpose, burial or procession. But a circle? Sometimes I have the feeling that they were erected simply out of a desire to baffle others – a practical joke, or an installation intended to reveal just how credulous we can be. Take Stonehenge. To Geoffrey of Monmouth, Arthurian myth-maker, the stones were Arthurian, brought from Ireland and assembled by Merlin. The seventeenth-century poet Edmund Bolton – whose greatest ambition was to write a history of Ancient Britain – claimed they were erected over the grave of Boudicca. To Inigo Jones – classical revivalist – they were classical, a temple to Coelus and a celebration of the symmetry which underpins the entire universe. To Walter Charleton – from the resurgent Stuart court of the Restoration – the stones were about monarchy, a coronation site for the ancient kings of the Danes. William Stukeley – self-styled Druid – saw them as Druidic. The essayist Horace Walpole was among the first to spot the pattern: 'whoever has treated of [Stonehenge] has bestowed it on whatever class of antiquity he was particularly fond of'.

In the nineteenth century, as the number of known circles grew, so did the variety of ideas about them. Stonehenge was a burial site. It was a court, a temple, a memorial. It was an observatory to worship the sun, an ancient calendar. It was an example of 'the psychic unity of man'. It revealed primitive endeavour, low-down on the slope that would one day lead all cultures to the uplands of civilisation. Diffusionists and invasionists fitted the stones to their own theories. Freudians saw a phallic cult. The cacophony of interpretations, the wild conjectures of counter-rationalists and mystics, led the donnish archaeologists of the mid-twentieth century to stuff their fingers in their ears. In their own work, only measurement, typology and comparison were allowed. That left the field wide open – and, in the 1960s and 1970s, Stonehenge became a hub of leys and energy-lines, a Neolithic computer, a portal for alien spaceships and UFOs.

And now, in our own pluralistic age, all-comers jostle at the cordons around the stones. Latter-day Druids are permitted entry for solstice rites, while paying visitors can assemble their own mix of theories like choosing souvenirs in the English Heritage shop. Radiocarbon dating and pollen analysis have added their precision to the debate, creating new chronologies. Post-holes beneath the car park suggest a large wooden structure that dates back to the eighth or ninth millennium BC.

The truth about stone circles is that you can make of them what you will. Interpretation reveals much more about 'us' looking than about 'them' building. The *why* is not the point. Perhaps the best way to understand the stone monuments of the late Neolithic is not by assessing their function in ancient society but by seeing them as works of art, long-ago versions of our perennial response to the beauty and mystery of existence. (Or maybe that's just me.)

In Armenia, there's a 'stonehenge', too – at Zorats Kar in the southern highlands. Another name for the site, Karahunj, literally means 'rock-henge'. I once spent a night there in the late 1990s with my friend Tigran, a film-maker. We took lavash bread and cheese and a bottle of Armenian brandy, and in the evening the shepherds came and sat with us. They had no more interest in the stones than their sheep, nosing and nibbling all around us. We talked of the war in Karabakh and the crooks in Yerevan, toasted the mountains and, with the brandy gone, they followed their flocks down into the valley. In the brick-coloured dusk, Tigran and I explored the site again, comparing our readings. The stones were very ancient – perhaps sixth millennium BC – twice as old as the sarsens at Stonehenge. They were the signifiers of a vast necropolis. They were an observatory. The strange apertures chiselled in them were aligned to certain stars and planets. They were, by any measure, extraordinary.

Later, inside sleeping bags, propped against the stones, we gazed up at the velvety dome of sky. Tigran told me of his mounting political rage and we laughed when he saw in the stars the outline of a raised fist. Later, off to the east, I spotted my own shape: a baby lying on its back, one leg raised; in six months' time I was due to be a father for the first time.

The next day, the fog had cleared. I had stayed inland but, walking north and west, I could detect in the underside of the clouds a reflection of the sea. At Sancreed holy well was another cloutie-tree, another rag-rug of threads and notes hanging from the willow. Squeezing down the steps into the well, I saw in the shadows a patch of tiny green lights: goblin's gold, or cavern moss, *schistostega pennata* – its lens-shaped cells have the capacity to amplify the faintest of light. The only other place I have seen this plant was at the Iron Age village at Carn Euny, in the mysterious subterranean chamber of the fogou. Might sanctity or strangeness be a condition for the plant's germination?

In the slow heat of the afternoon, I found a wall in a deserted valley and leaned against it to read and generally look. A vixen came padding up the valley. I froze, and watched her, half-hidden as I was behind the ferns. She ambled past me just two metres away, and I could see her hanging tongue glisten in the heat. She climbed the wall and, before jumping down the other side, looked back over her shoulder, straight at me, as if to say: 'I knew you were there all along. So what?'

I was nearing the coast; the land was becoming more marginal. I came across an abandoned farmstead. On one side was a cottage without its roof; the main dwelling looked almost intact. The door creaked open on damp-stiffened hinges and I stepped into a single large room. It had the dusty, fungal smell of age. An old Cornish range stood busily rusting to one side; from the chimney above, decades of soot dropped to a delta-shaped mound on the hob.

I pressed the bottom stair with my boot. It groaned and sagged, but held. Something was moving upstairs. I climbed, step by step. As my eyes came level with the floor, I saw scattered plaster on the bare boards, streaks of guano on the walls and suddenly – *whoosh*! A barn owl. Its wings filled the room. It looked monstrous. Then it was gone – out through a broken window, leaving a lone feather zigzagging down to the floor. In the silence, I could hear only the beating of my heart.

Up on the moor that day, everything shone bright and clear. Water stood in steel-blue puddles across the paths. In the distance was the skyline profile of Chun Quoit and I headed over to it. Such quoits are said to resemble giant mushrooms – their vast capstones propped on a stalk-like base. In which case, Chun is a fat boletus.

Comparison can also be made between the quoits or cromlechs and natural rock formations like the tors. On his way to Land's End in 1860, Francis Palgrave recalled: 'I was struck…by the likeness between the masses of rock, piled up by Nature only, and those cromlechs which also occur in Cornwall.' More recently, in an article entitled 'An Archaeology of Supernatural Places: The Case of West Penwith', Christopher Tilley and Wayne Bennett used the similarity to develop the notion of imitation: 'In the Neolithic, people were making an effort to replicate ancestral work, erecting massive stones, attempting to mimic the tors.' The perception of the tors as man-made, goes their argument, led settlers in the Neolithic period to build something similar and in that way procure ancestral licence to use the land. It is an attractive idea: doing the 'right thing', adopting local traditions, confers belonging on incomers. I had been aware of this during the renovations at Ardevora. But I couldn't help thinking that a part of the theory's appeal comes from our own attempt to rediscover a sense of belonging to the land.

Reaching the coast in the early evening, I watched through binoculars a climber spidering up the granite face of Bosigran cliff, dipping into his waist-band chalk bag before each move. The cliff beneath him dropped sheer to a pool of turquoise where slow swells slopped against its base. I carried on, into the bright evening. Gannets flew westwards along the coast below me, in groups of threes and fours. They followed the shoreline with their steady, determined flight, down from the colony in Pembrokeshire, heading to the fish-filled waters off Land's End. The forecast was better now for some days ahead – did they *know*?

The coastline stretched off into a pale haze; the seas below soaked the cliffs with surf and the sun was dropping, its light flooding across the water, making it look like nowhere on earth. It was a strange sensation, feeling such a bulk of land at my back and everything ahead dissolving. I'd started the day early and had now clocked up twelve hours in the open air; the brightness sat heavy on my brow and my vision was jumpy. In the 1960s, the guide-book writer M. A. Hollingham recalled the 'altered perception' he experienced walking this piece of coast: 'We began to feel that we could walk like this for ever, never having to turn back but always going on, to see what was round the next headland, and the next.' But he soon found that the place made him want to cry out: 'Stop! It's all too much, we can't take any more.'

Charles Dickens also felt a sudden intensity on approaching Land's End. He and his friends had had a riotous time coming through Cornwall – 'I never laughed in my life as I did on this journey' – but arriving at these last cliffs, they all turned sombre. They saw the sun sink into the Atlantic, and 'each in his turn declared [it] to have no parallel in memory'.

Many who have loved the area around Land's End have fallen foul of its darker moods. John Thomas Blight was seven when Dickens and his party came to a halt on the edge of the ocean. Son of a Penzance schoolmaster, Blight had already learnt the rewards to be had from roaming Penwith's moors, scrambling over granite cliffs, squeezing into shore-caves, swimming in tide-pools. In time, his boyhood adventures grew into more profound study. No one except William Borlase has ever spent so much time here in West Penwith exploring, writing, botanising, and surveying its ancient monuments. He recorded the area's rare plants, birds and fish. He began to paint – delicate still-lives and day-to-day scenes from along the coast. His work saw him elected a Fellow of the Society of Antiquaries. At the same time he produced a guide book, *A Week at the Land's End,* aimed at the growing number of visitors who stepped off the train at Penzance's terminus. In it he promised 'scenes and memories which no other district of our land affords'.

By the mid-1860s, Blight was assembling material for his most ambitious project. Borlase's great survey needed updating; the understanding of antiquity had moved on. In *The Cromlechs of Cornwall,* Blight drew widely on the growing body of ethnographic literature from all over Europe and beyond. He aimed to place west Cornwall's monuments in a universal context, as an example of the perennial need to commemorate, to erect in the landscape stone markers to offset our fleeting presence on the earth.

It proved too much. 'I have a notion, of which I cannot be rid,' he wrote to a friend at this time, 'that my brain is all wrong.' He was finding it harder and harder to concentrate; he became infatuated with a woman named Evelina Pidwell and grew convinced that the Anglican clergy had been infiltrated by Druids, who were just biding their time before reintroducing human sacrifice.

In the archive of the Morrab is the only existing copy of *The Cromlechs of Cornwall,* the uncorrected proof which he was going through when he fell ill. The margin of the last worked-on page is a crazy glut of added commas, as if each word needed protecting from the next. The book was never completed. Blight slipped from public view. A decade later, his publisher announced that he had died: 'daily toil brought on the illness which had so sad a termination.'

In the hall of the Morrab Library, tucked away beside a light switch, is a photograph of Blight in his mid-thirties. Some of the black in the clothes has been transformed by the years, reverting to a sheen of silver nitrate. His big eyes, long hair and bushy whiskers reminded me of Neil Young from the *Sugar Mountain* album sleeve. But Blight was already very unwell by this time, and beneath his eyes hang crescent shadows.

Explaining the mental collapse of anyone always seems inadequate;

it is made no easier by the separation of a century and a half. But the impression remains that Blight, like John Leland before him, was broken by the sheer size of the questions he had taken on – of life and death, land and the past. There have been times writing this book, trying to reach the meaning of a place across the ages, when I have felt a shadow pass over my desk. At other times come those episodes of intense connectivity, when notions zing out along the threads of an ever-expanding web, and colour everything with their hints of absolute truth. Certain subjects have the capacity to swallow you whole and one of these is the interpretation of ancient monuments.

In fact, ten years after his disappearance, Blight was still alive. In May 1871, escorted by two guards, he had been taken from Penzance to St Lawrence's Asylum in Bodmin. Two notebooks survive from his first months there. Reading them in the basement of the Morrab Library was almost unbearably poignant. The writing is microscopic. Bending over the pages, I felt I was peering down on a colony of black-ink insects.

Blight was sharing a room with nine other men. He writes of their foibles – Captain Hill reading Job, saying to no one in particular, 'Do you want any money?' and singing loudly in his 'seaman voice': '*Mary sat weeping. Mary sat weeping! She knows not where he lay . . .*' A bird flies around their room, and Blight watches it, recording in meticulous detail its landing on a gas lamp and its circling, circling. 'May that bird live for ever,' he jots down in hope.

There are ramblings, pages and pages about Rome and Jerusalem and Evelina Pidwell and the wickedness of man. But among it, too, are more coherent thoughts on antiquity and the work he left unfinished. At one point he predicts our own current understanding of funerary monuments: 'graves were often purposely placed near or in a line with some remarkable natural object to add importance to the mounds'.

Confusion and lucidity – and the slow shrinking of his world. He sits on his bed. He looks at his sock. 'It's been repaired with brown wool, the main structure of the socks being a dull blue colour. I think it was last week I first found it repaired in that manner, but who so repaired it I do not yet know.' He draws it, the pattern of darning distinct from the original knit. The pages are dotted with drawings and little washes. I turned over one, and there – life-size – a wasp, so beautifully painted that I started when I first saw it. He'd been watching it crawl over the back of his hand.

John Blight stayed on at the asylum for forty years. He died there in 1911. Only a few people knew he was even alive. He never went west again, to his beloved Land's End. On occasions he was allowed out for the day. In the summer of 1900, there was a dig at Harlyn Bay and a chambered tomb was discovered. A photo catches a group of hatted men and women gathered around the excavations. To one side of the group is a small, white-haired man. He holds a pencil over a piece of paper, and stares down at something one of the diggers has retrieved from the grave. This man has been identified as Blight. And here in his notebook is the sketch from the same day, of a skeleton.

'Fancy my friend who reads this,' the forgotten Blight wrote in one of the early asylum notebooks, 'friend or enemy, the quiet of those urns left in some wilderness, common, moor, tarn field, plain, rock, or cairn.' He imagines the urn disturbed by some passing animal and the human remains within it crying out in the emptiness: 'Aye! I be! Aye! I be!'

It was almost dusk when Pendeen lighthouse came into view, squat and cruet-shaped. Here the coast takes a sharp left turn, and heads south towards Land's End. Out at sea, the wind was kicking up short white-topped waves where the ebb tide was rounding the point. The horizon ahead was a yellowy fusion of sea and sky, and the long low clouds hung there like a scattering of rocks.

I stayed that night in a pub called the Radjel (*radjel* is one of several Cornish words for a 'pile of stones'). The window looked along the main street of Pendeen, itself little more than a strip of houses on each side of the main road. I met a man in the bar who'd worked for years in the merchant navy. He'd retired here not for its sea views – 'I'd be quite happy never to set eyes on it again' – but for its geology: 'Greenstone and hornfels and granite – ground's like a big fruit cake.'

In the morning, I dropped down to the coast and entered a mile or so of almost continuous mine-scar – ranges of orangey spoil-heaps, old spalling floors, stamp yards. At Levant, a wreath was hooked on a stake, hanging down over the cliff. The heather flowers had turned to dust and the red ribbon was rain-washed pink. Far below – almost in the water – was the former engine house. The shaft had pushed out beneath the sea bed for a quarter of a mile. During gales, the miners would hear – and feel – the Atlantic swells booming against the rocks above them. In 1919, thirty-one men were killed when the lift-cable snapped and they plunged down the shaft.

At Cape Cornwall the slopes of the golf course looked less like grass than the brushed cotton of a green blouse. A path ran alongside it, up from Priest's Cove, and at the top was suddenly the great sweep of Whitesand Bay, the glitter of the sun on the wave-backs – and beyond it, on the skyline, the distant cluster of buildings at Land's End.

# 17 | NANJIZAL

From Cornish *nans-*, 'valley', and *-ysel*, 'low', in this sense probably referring to the valley's deep-cut sides.

LAND'S END – IN CORNISH, *Pedn Wlas* (literally 'end of the land'). *Finistère* in France (the name in Breton is *Penn ar Bed*, 'end of the world'), *Finisterre* in Spain (*Fisterra* in Galician). It is as if the farthest prod of rock into the ocean presents a topographic fact so fundamental that it can only be described in the most direct way. But these westerly capes carry a mythology as weighty as any place in Europe.

Some years ago, I spent a few days making the pilgrimage from Santiago de Compostela down to Finisterre. As a ritual destination, the headland predates the Christian pilgrimage route by hundreds, and perhaps thousands, of years. I did not go there direct, but via La Coruña and along the stretch of coastline known in Galician as the *Costa da Morte*, the 'Coast of Death'. A hundred and forty-eight ships are known to have been wrecked in this area; countless other losses have gone unrecorded, the open boats and the *percebo*-hunters washed from the rocks. Above a deserted strand, I came across the Cementerio de los Inglesos, where one hundred and seventy-five British sailors from the *Serpent* are buried. In a local restaurant, I saw a plate decorated with an English hunting scene; on the back was stencilled: *Serpent*. A man in Muxia showed me his own private museum of wreck-booty – brass binnacles, barnacled timber and, oddly, the first telephone directory for the city of Liverpool.

That's the popular understanding of the name, the Costa da Morte: a coast of shipwrecks. But there is another explanation, one linked to the Finisterre pilgrimage. In pre-Christian Europe there was a widespread belief that when the sun dropped below the horizon, it went on to shine on the realm of the dead. The belief gave rise to – or arose from – the tradition of imaginary places beyond west-facing shores, like the Isles of the Blessed and Tir na nOg. The passage of departed souls followed the sun's course across the ocean. To stand at the limit of the land, to witness in life the sun's slide into the sea, was understood to be a rehearsal for this last perilous journey.

At Finistère's Baie des Trépassés – 'the Bay of Souls' – holy men used to sail out to the Ile de Sein to enact entrance to the next world. It was believed in Galicia that pilgrims who made their way to the headland of Finisterre were guided by the Milky Way, which itself was made up of dead souls gliding towards paradise. Rather than suppress such a tradition, the Church re-invented it, transferring it a little inland, and involving a miraculous field of stars – the '*compostellae*' and a shrine for the bones of St James. The Milky Way then became known as the Way of St James. *La Romera de Santiago*, a drama of the Spanish Golden Age, contains the lines 'the pathway in the sky, because it is filled with stars / was called by the pagans / The Milky Way, / commonly known / as the way of St James'. The scene involves a woman taking the marital bed outside and her husband warning her that if one of the pilgrims above should drop anything, it would fall and 'break their heads'. They think about that for a little bit, then move the bed back inside.

A robust cult of the dead has long survived in north-west Galicia. As part of his five-year tramp around Spain in the 1830s – 'circulating the scriptures' – George Borrow was caught out on a wild Galician moor. His guide became anxious, afraid that in the coming darkness they would encounter the *Estadea*, the assembly of the souls. 'I tell you frankly, my master, that if we meet the assembly of souls, I shall leave you at once, and then I shall run and run till I drown myself in the sea.'

In the port of La Coruña, I spent an afternoon with the writer Manuel Rivas. 'The dead are always with us in Galicia,' he explained, 'often in surprising ways.' He had recently published *The Carpenter's*

*Pencil*, a novel set in Santiago during the Spanish Civil War. The story features an executed prisoner who is shot by a guard, only to take up position in a pencil behind the guard's ear. Once you permit the logical re-emergence of the dead, anything is possible, and Galician folk tradition is enriched by the mythology of Fisterra. In the work of Gabriel García Márquez, Manuel recognised the same porous reality, where the fantastic and the mundane mix freely. 'Márquez's grandmother was Galician,' he explained. 'He said that a lot of that magic realism came from listening to her stories.'

Over the next few days, I watched the sun sink each evening into the sea, and at night saw the Milky Way follow a similar path. Walking the final couple of miles to the end of the land at Cabo Fisterra, I spotted an odd group of concrete structures built into the cliff. They looked like outsize bookshelves, without the books. A man was sitting on the verge beside his bicycle.

'What are they?' I asked.

'Them? They are *las casas das mortas*.'

'But they are empty?'

'Empty, yes. But people have bought them, people from Madrid. A famous architect designed them. When they die, these rich people want to be buried here, in the boxes. I don't know why.'

At the far end of the mile-long Whitesand Bay, the headland of Land's End stood bright in the sunlight. But it grew no closer. The sand beneath my boots sunk with each step. It felt like trying to run in a dream. I stopped to watch a surfer paddle in for a wave. He popped up and slid along its face, turned once and then flicked his board back behind the crest. I saw him reappear, already prone, heading out again. I hobbled on in my walking gear.

Above Sennen the path climbed again to the cliff. After the sand, here was geology again: bare rocks worn into a gallery of shape and suggestion, a cliff castle and a sweeping saddle beyond it, and, on the rocks below, a casualty: the rusting remains of a ship.

And then, Land's End itself. First came a fenced plot of sand and slides and zip wires, a model pirate ship and the old Padstow Severn Series lifeboat, high and dry in the WRECKREATION AREA. The main entrance to the complex of white, rough-render buildings was beneath a grand arch. Six Tuscan columns propped up a slate-hung pediment below which was written: LAND'S END. Inside, the courtyard was cleverly protected both from the weather and from the view. Instead hoardings advertised entertainment – *Arthur's Quest Experience: The Legend, 4D Film Experience – Skull Rock*. This was not so much Land's End as the Land's End Experience.

Beyond the shop and theatres and the hotel, and beyond the staff car park, was nothing but rock and sea and sky. Above the cliffs it said: DANGER – CLIFFS but it wasn't difficult to find a way down between the wind-smoothed granite blocks with their sprigs of sea ivory, down a gully and out on to a ledge so thickly grassed that it felt like a sofa, there to sit back and gaze out on the sea.

Little but this has ever happened here: no invasion, no nation-stirring oratory, no rock-residency of a silver-tongued monk. So where does it spring from – the urge to reach this point? The area's concentration of monuments from prehistory leaves the vaguest hint of some ancient belief concurrent with that at Finisterre – of the passage of souls out over the sea, to a place of limbo far beyond the horizon. The only other possible sign comes in the name of the neighbouring parish of Morvah, or 'sea-grave' in Cornish.

The headland's current popularity is recent, and it tells the wider story of how in the last few centuries topography has seeped into popular consciousness; not local topography based on what you can see, but map topography – that abstracted version of space which creates its own imperatives. The earliest surviving accounts make nothing of Land's End. John Leland went just about everywhere; he wrote extensively about Penwith – but he didn't come to the headland. With the Reformation came a growing awareness of the land's wider form, as well as a number of topographic books that drew on Leland's example. In 1602 Richard Carew published one of the most extensive: *Survey of Cornwall.* Reaching the end of his account, here in the far west, he illustrated the land's tapering with a shape poem:

> I will refresh you, who have vouchsafed to travaile in the rugged and
> Wearysome path of mine ill-pleasing stile, that now your journey
> Endeth with the land; to whose Promontory (by *Pomp.*
> *Mela,* called *Bolerium*: by *Diodorus, Velerium*: by
> *Volaterane, Helenium*: The *Cornish, Pedn*
> *an Laaz*: and by the English, The
> lands end) because we are
> arriued, I will heere
> sit mee downe
> and rest.

By the mid-eighteenth century, touring was widespread. Criss-crossing the country, seeking out all that was sublime or picturesque, the growing tribe of lettered travellers carried a new sense of the nation's curves and extremities. Those venturing south-west followed the instinctive logic of reaching the peninsula's end. A couple of place names here at Land's End itself are specific to this time: a mossy boulder resembling a man in a periwig is Dr Johnson's Head, while a small promontory just to the north of Land's End is Dr Syntax's Head, after the main character in William Combe's satire *Tour of Dr Syntax in Search of the Picturesque.*

By the mid-nineteenth century, as first-hand accounts of Land's

End grew in number, visitors brought ever greater preconceptions. Following one of his epic bouts of walking, George Borrow opened his notebook here on 20 January 1854: 'Wrote this at the Land's End... Nothing could be more grand than the scene, the sea bursting horribly against the battered rocks.' Ten years later, the American Elihu Burritt reached Land's End from London on foot: 'Here I was at last, at the extreme southern end of England! It was reached and realised – the dream of twenty years agone!' As she approached Land's End, the novelist Dinah Mulock Craik found herself similarly excited: 'We wondered for the last time, as we had wondered for half a century, what the Land's End would be like.' On seeing it at last, she gushed: 'not to have stood for one grand ten minutes where in all our lives we may never stand again, at the furthest point where footing is possible, gazing out on that magnificent circle of sea'.

She was not alone. W. H. Hudson has left an account of something he witnessed on the morning of 24 May 1907. He was at Penzance station when a train from Manchester pulled in. Several hundred people spilled from its carriages. By midday, another three such trains had arrived. More than a thousand people had now moved along the platform, out into the station's concourse and on to the harbourside. They had a mere seven hours before the trains would leave again, for the fourteen-hour journey back north. They had one object in mind – to reach Land's End. Dozens of carts, charabancs and motor-buses shuttled them the dozen miles to the farthest point of the peninsula; many did not make it. The Cornishmen were baffled. One said that he'd lived here all his life, within ten miles of Land's End and had never gone there, never even wanted to.

Whether these visitors were re-enacting a long-forgotten tradition is now hard to tell. In any case, perhaps it's something intrinsic. Mountains house the gods, caves are entrances to the underworld, sacred rivers wash away sins, and ferry crossings are places of transition. The western extremities of the Eurasian landmass have their own cosmographic suggestiveness: the sun sliding out above the ocean and over the empty horizon is entering the realm of the future and the time beyond our own death, and because we seem incapable of accepting voids, we fill the waters there with supposition, with islands of eternal life.

W. H. Hudson himself was curious about Land's End. Brought up in Patagonia, he came to England with a long-held desire to reach its south-western extremity, of which, he wrote, 'a vast and misty picture had formed in childhood'. One cold day, early in that spring of 1907, he arrived at Land's End and found there a group of old men. They were each sitting on separate boulders, in the sun, wearing scarves and greatcoats. Some were resting their closed palms on the tops of walking-sticks, some had their chins in their hands. They had come, he discovered, from all over the country.

Hudson imagined each one of the old men privy to a vision, his fuzzy eyesight suddenly replaced with a final fierceness and clarity, each making his own imaginary voyage out over the water to a group of 'uncharted' islands. There he would discover not 'a beautiful blessed land bright with fadeless flowers, nor a great multitude of people in shining garments and garlands who will come down to the shore to welcome him with sounds of shouting and singing and playing on instruments'. Instead, Hudson speculates, it is the very opposite of the sort of place idealised in life, of dappled shade, sparkling water and green fertility, the *locus amoenus*. It is free from the agitation of growth, without movement, empty, 'a silent land of rest':

> He will land and go up alone into that empty and solitary place, a still grey wilderness extending inland and upward hundreds of leagues, an immeasurable distance, into infinity… And when he

has travelled many, many leagues and has found it – a spot not too sunny nor too deeply shaded, where the old, fallen dead leaves and dry moss have formed a thick soft couch to recline on and a grey exposed root winding over the earth offers a rest to his back – there at length he will settle himself. There he will remain motionless and contented for ever in that remote desert land in which is no sound of singing bird nor of running water nor of rain or wind in the grey ancient trees.

*

It was early evening. I climbed back up the cliff. To the south of Land's End is a stretch of coastline as remote and dramatic as any in Britain. Dark cubes of granite balance on the headlands; there are high rock-arches, caves, blow-holes, chimneys and towers, and, in amidst them all, stretching across a bay like a smile – a beach. Dropping down to it, I took off my boots and wriggled my feet down so they almost disappeared. Picking up a handful of sand, I pushed it around my palm and saw it was made up of broken shell, tiny chips of cream and yellow chiton. I went for a swim. The beach was bouldery and in the surf I struggled to stay upright, until far enough out to kick away and transfer my weight to the water. I let a couple of breakers tumble me. Back on the sand, I walked out to where a high cave opened to the sea. In the yellow blaze of evening, I stepped into its cathedral space and listened to the rush of water, surging in over the rocks to bubble at my feet.

In a grassy hollow above the beach, I pitched my tent. During the night, the wind freshened and the tent flapped like an unsheeted sail. Twice I went out to secure the guys and the second time I stood for a while on the cliff edge watching the silver wash of moon on the water and feeling like the last person alive. Dawn was stark. I crawled out for the last time and dressed. Small clouds were dashing across a sky still spotted with stars. The wind funnelled down the valley and out over the beach. Gannets were feeding beyond the breaking crests. They glided back and forth on wide wings, and without warning would drop into the water. It was easy to miss them. I watched one pushing south with a determined flight. I was sure it was too set on its way to waste time feeding – but then, the faintest pause, a dip of the head, a tuck of the wings, and it was diving, piercing the water like an arrow. It stayed down for a long time. Then it was back on the surface again, shaking itself, rising again. Closer in, the waves lined up in long ridges. They darkened as they approached the shore and the under-roll slowed with the rocks on the sea bed. As each top turned and broke, the wind caught the lip and flicked it up in a mane of back-spray. I knew that place from surfing, when the offshore wind blows the white water off the top and it falls in the trough behind with a noise like a rain shower.

A yacht was crossing the water outside the bay, running north under jib alone, hurrying around the dangerous cape. Far off in the haze I fancied I saw the faint line of the Isles of Scilly – but I may have imagined it.

# 18 | LETHOWSOW

Possibly from Middle Cornish *lethas* (plural *lethowsow*), 'milkiness', referring perhaps to the appearance of surf on the Seven Stones reef. Carew (1602) uses this form.

Beyond land's end, in many fathoms of water, lies a ruined settlement called simply the 'Town'. It was large enough in its time for the streets and back-lanes to contain a hundred and forty churches. The Town was part of a bigger tract of land known in Cornish as Lethowsow, and in English as Lions or Lyonesse. But one day, the seas rose and flooded the Town and swamped the forests nearby and drowned everyone there. Everyone except one man – a single horseman, who managed to gallop ahead of the surging waters. Such was the fear of the sea rising again that for many years the Vyvyan family at Trelowarren on the Lizard kept a horse in the stables, permanently saddled. There are stories of fishermen hauling up doors and windows in their nets and, when the wind is still and the sea smooth, some have told of hearing from under the water the faint sound of ringing bells.

Such mythical floodings have afflicted other places – from convents in Poland, to whole towns in India and Ethiopia. But Cornwall has proved particularly adept at sustaining the story. Lyonesse is referred to in Geoffrey of Monmouth and in Wace's *Roman du Brut*, and in the cycles of Tristan and Iseult; rumours of lost land off Cornwall are mentioned in William of Worcester's *Itinerarium* and by John Leland. Richard Carew in 1602 put Lyonesse firmly on the map, elaborating it in his *Survey of Cornwall*: 'Lastly, the encroaching sea hath ravined from [Cornwall] the whole country of Lyonesse, together with divers other parts of no little circuit.' Carew had already told William Camden about Lyonesse and its inclusion in Camden's *Britannia* helped carry it on through the centuries, adding to the cartography of places which, unable to be visited, become repositories of our deepest feelings. Lyonesse held the idea of loss – of that vanished era when everything was larger and better. By 1922, Walter de la Mare could write: 'In sea-cold Lyonesse,/ When the Sabbath eve shafts down/ On the roofs, walls, belfries/ Of the foundered town'. Two decades later, in *Brideshead Revisited,* Evelyn Waugh has Charles Ryder lamenting his years at Oxford: 'submerged now and obliterated, irrecoverable as Lyonnesse'.

But it was Alfred Lord Tennyson who did most to raise Lyonesse. In 1833, he embellished the stories of flooded ground and wrote of a city and a mountain that had 'topple[d] into the abyss'. The mountain was 'the most beautiful in the world, sometimes green and fresh in the beam of morning, sometimes all one splendour, folded in the mists of the West'. Beauty was all well and good but the real success of Tennyson's branding was the addition of King Arthur. The mountain that rose from the land of Lyonesse was called Camelot, and Tennyson placed it beyond Land's End where, 'save the Isles of Scilly, all is now wild sea'.

Later, in the *Idylls of the King,* Tennyson went further. Lyonesse was the site of Arthur's last day on this earth. The ancient land rises once more out of the abyss, and in thick mist Arthur fights his final battle, 'the weird battle in the west'. Afterwards, fatally wounded, he staggers over the resurrected ground. All his knights except Sir Bedivere are dead. No sorrier figure exists in English literature than the dying Arthur – the disillusioned king, betrayed and bereaved, with nothing to show for all his efforts and high ideals, living out his last hours in the sunset regions of the far west. Bedivere carries him to the land's end, where the barge takes him away into the western ocean, to a calmer place, a *locus amoenus*:

> Where falls not hail, or rain, or any snow,
> Nor ever wind blows loudly; but it lies

Deep-meadow'd, happy, fair with orchard lawns
And bowery hollows crown'd with summer sea.

# 19 | SCILLY

Early forms – Sulling, 1160, Sully, 1176 – suggest a possible link with the pre-Christian goddess Sulis. The Norse *-ey,* 'island', might have been added to the goddess's name. The 'c' in modern spelling was added to prevent too close an association with the English 'silly'.

On the eastern edge of the Isles of Scilly, the last of those volcanic intrusions that form the uplands of the south-west, are the uninhabited Eastern Isles. Some are no more than surf-rounded rocks, breeding sites for gulls and fulmars, stands for wing-drying cormorants and shags. Others are islands thick with a rare species of bramble (*rubus ulmifolius*) that binds with the gorse to cover the ground like a thorny pelt. One of these islands goes by the name of Arthur, pronounced by Scillonians 'ar-ter', the old Latin way – an indication, it is said, of the place name's antiquity. Three entrance graves stand high on the ridge of Arthur.

I took the *SS Scillonian III* to the Isles of Scilly and for a week stayed in a clapboard hut on St Martin's, the closest inhabited island to Arthur. The weather was terrible, with a high southerly swell that prevented any chance of reaching the Eastern Isles. The two-room hut stood among high hedges of euonymus and pittosporum. It was painted powder-blue outside, and had a felt roof that purred when it rained – which it did, a lot. Each morning a neighbouring couple dropped by to say hello. They were inseparable and at first light, sitting at the table with my coffee and books, I could hear them coming along the path: '*WAK-WAK... orr-wooor-wooor... WAK-WAK...*' They were an elderly mallard drake and a speckled hen.

Scilly is baffling: like all the best places, it draws you into a state of ceaseless questioning. How does it all work – how do people run their lives? What would it be like to be here for a long time, to go through the winter? But it's not just lifestyle curiosity. The reduced scale brings bigger questions closer, and adds a heady combination of freedom and isolation. John Donne was wrong when he wrote that 'No man is an island'. We are all islands. The Polish poet Zbigniew Herbert came closer to the truth in his *Elegy of Fortinbras*:

> It is not for us to greet each other or bid farewell we live
> on archipelagos
> And that water these words what can they do what can
> they do prince

John Fowles, whose mother was Cornish, wrote an extended essay based on his lifelong fascination with Scilly. He too found in the physical form of islands a version of ourselves: 'It is the boundedness of the smaller island, encompassable in a glance, walkable in one day, that relates it to the human body closer than any other conformation of land.'

On islands, time stretches out, footsteps lighten and everything real becomes partially suspended. Scilly also has another dimension to it – the underwater. The central cluster of islands are merely the high points of what was once one big island – and not in some unpeopled geological era, but within the span of written history. Two Roman sources – Solinus and Sulpicius Severus – refer to the 'Scilly Isle'. The main group of St Mary's, Tresco, Bryher, St Martin's and the Eastern Isles were hills connected by a flat area of moor and forest in the middle, where now there is shallow sea. Just below my cabin, towards the dunes, grew a tiny group of wood spurge, indicator of ancient woodland. Whenever I passed it, I stopped – filled with affection for the little plants, their rarity and survival.

On my first full day on the island, I looked up Keith Low – crofter, fisherman and amateur archaeologist. I'd met Keith here once many years ago and now I found him down on the beach in his sun-faded Breton cap and red neckerchief. He was tending to his small boat and grumbling – like all good fishermen – about the weather. In the bottom of his punt was a plastic laundry basket with some crabs from his keep-pot. He took a small one and chucked it back into the water. 'Leave it, Inca,' he shouted to his dog. 'LEAVE IT!'

He vaguely remembered me. I said we'd spent a day fishing round the back of the Eastern Isles and he'd opened his arms at the whole group of uninhabited islands and announced: 'All this, Philip – this is my office!'

He chuckled. 'Sounds about right.'

'I'd like to try and reach Arthur,' I said.

'Arter?'

'*Arter.*'

He took his bowline from the boat, and clipped it to the outhaul. 'If you've ever done any divining, with the rods, Arter'd drive you mad. I can try and take you over there – soon as this damn weather improves.'

We walked up the beach. 'Let me show you something.' Towards the top of the sand were two rocks. As soon as he pointed them out, I could see they'd been deliberately placed. Keith gestured beyond them, in each direction. 'Stone row – used to be a whole line of them here. Now you don't build a row of stones on a beach, do you? Would have been high and dry when they did it.'

After a few days on Scilly, you develop an awareness not just of a submerged landscape, but the invisible contours of a ritual landscape. The islands have an astonishing number of sacred monuments from the late Neolithic and early Bronze Age. The building of them went on longer here than on the mainland, perhaps until as late as the seventh century BC. Their concentration is greater on Scilly than anywhere else in the country – and that's just those above the present sea level. Looking around at the skyline, at all the saddle-back islands and the rocks, I kept thinking of it as a flooded version of Bodmin Moor, where all but the top few hundred metres were under the sea.

Keith lived in a cottage in Higher Town and his front garden was crammed with fishing gear – parlour-pots and cobles, coiled warps and cane-and-flag marker buoys. Inside, every surface was covered in books. 'Last winter's reading,' he said. 'Haven't had the chance to clear it away.' The spines showed an omnivorous diet of history, political biography, fiction and natural history. He pointed upstairs. 'There's another fifteen thousand up there.'

A couple of friends had come to visit him from the mainland, and they'd all been cockling on St Martin's flats, the submerged plain on the island's western shore. We sat shelling and talking and drinking. Keith had a lovely raconteur's voice and told stories with the unhurried pleasure of those who are not always thinking they need to be somewhere else. On a table was a photo of him in the same Breton cap, the same red neckerchief. He is standing by the gate of his cottage, yarning with the same slow ease. A man is looking up at him respectfully. 'Just chatting with the landlord,' muttered Keith. It was Prince Charles.

Keith then explained how, when he was seven, he'd visited a dig on the neighbouring island of Tean and the archaeologist there had stopped to talk to Keith and explain what was happening. It was a conversation of a few moments only, but one that triggered a whole lifetime's interest in uncovering the past.

Later we went outside and looked out over the end of the island. The cottage had a direct view of Arthur. The sea was quiet on this side, in the island's lee, but to seaward, I could see the water still white off the western rocks.

'Wind's due to drop away tomorrow night,' mused Keith. 'Maybe we'll get out the day after that. And when we come in to Arthur, I'll give you the helm. Make you see better.'

I knew what he meant: as soon as you're holding the helm of a boat, your gaze sharpens.

It blew for another two days, then I woke in the hut and for the first time heard no wind in the eaves. In the silence, still some way off, I could make out '*Orr-woooor*... *WAK-WAK*.' I phoned Keith and he said he'd be checking his pots that afternoon; he could drop me off on Arthur. That left the morning for something else that the weather had prevented. With the seas settled and the tide out, I hurried over to the dive school to pick up the snorkelling gear I'd hired. At Wra Ledge, I waded into the shallows, and swam out towards Guther's. Rinsing the mask, I pulled it down over my face, chewed on the snorkel, blew out, then dipped my head.

The upper world disappeared. An entirely new one took its place, full of drift and silence. I could hear only the bellow of my own breathing. Beneath me, with the tide still on the ebb, rolled a multi-coloured traffic of broken kelp and sea lettuce and red algae. A holdfast like a Viking club was being tugged along the sand as if by an invisible string. It was odd to think of this as dry land as recently as the sixteenth century. I followed a cuckoo wrasse, picking out the oranges and blues of its flank – until it sensed me near and dashed off, leaving just a puff of sand. Kicking down, I stretched out my arms towards the seafloor. Thick veins of light flickered on the back of my hand. The sediments below were blond and glittery. I pushed my fingers down into them, and they came out black. I dug them in deeper and the water was suddenly full of it, like squid ink. It was peat.

Only then did the true sense of inversion take place, of being raised above solid ground, gliding over it. The place that came to mind, with the black soil, was the bogs below Rough Tor, where Richard and I had been thwarted on that crisp day so many months earlier.

*

'That's the place for bait, down in the peat,' Keith shouted over the whine of the outboard. 'Ragworm thrive on it!'

We were motoring out towards Arthur. I had the tiller in my hand, and could see the high ridge ahead. Keith stretched out his arm and pointed to the sites of the entrance graves on the twin tors. I nodded. Then he shifted towards me and pointed again, more precisely. On the ridge between the two peaks – tiny dots on the skyline – were a couple of standing stones. 'There's more there – a boulder wall – or a stone row!' I brought the boat in towards the rocky beach, handed Keith the helm and jumped ashore. He slewed it round and I watched him go, the bow raised high by his weight astern, the peak of his cap pointing out to the rest of the Eastern Isles, to his office.

Several oyster-catchers flew off as I climbed up over the rocks, their sharp calls full of pique at my intrusion. Arthur is almost three islands – three summits joined by pebble-fields which are surged over by the biggest seas. On the first granite outcrop of Great Arthur there were signs of a gulls' picnic – a scattering of fish bones in a stone hollow. The ground above was covered with bramble and honeysuckle, so thick that it was like walking through spiky snow. I reached the first of the main peaks and found the entrance grave. Out towards the second peak, four standing stones were visible and I hunted for others, without luck. At the third peak, the grave had the vast capstone shifted off the chamber. I squatted down. Inside was the same tight-knit bramble from the slopes, the same honey-suckle. Looking into it, open to the weather, I was struck by a sense of emptiness far greater than that of this one tomb.

The ridge on Arthur fits the pattern of hero-sites, those high places associated with Oliver Padel's 'pan-Brittonic figure of local wonderment'. On a timeline, the adoption of Arthur as a name here would probably come closer to our end than the point thousands of years ago when the ridge was first deemed significant, or had its first monument built on it, or the first body carried up to it. Even in our own age of reason, places still have the capacity to fix the biggest of ideas, the most complex of mythologies, the most distant of pasts. Since the Neolithic, the land below these peaks has altered utterly. The seas now cover it. But those who sailed here from the mainland or worked the inner moors of the great isle would have looked to this hill and felt the same instinctive and layered attraction to it that visitors now have approaching more accessible places like Tintagel or Glastonbury.

I walked out to the southern tip of the island. Ranks of swell were queuing up to smash against the cliff. I picked one out, half a mile off, and followed it in. Today's westerly breeze was blowing across it, driving a crowd of smaller ripples obliquely to the larger wave, up over the rise and down into its trough. The face was now nearly sheer but still not breaking. As it came in close to the cliff, the water before it dropped away from the rocks below me, dragging rags of kelp into the foam below with a waterfall sound, and for a moment a new layer of barnacled granite was exposed, further down, a revelation of steps and gullies full of tumbling water. Then the wave broke and the rock I'd been watching was gone, a mass of seething foam.

One evening towards the end of that week, I took the last ferry out to Bryher – meaning 'island of hills' – and camped in a field below Shipman Head Downs. This end of Bryher is the most northwesterly tip of all the islands. If the ring of Scilly is Bodmin Moor submerged, then the top of Bryher is Rough Tor, and on Shipman Head Downs is the suggestion of a ritual site on the same remarkable scale.

I rose at dawn to see the cairn field. The skyline of Tresco, just half a mile across the water, was sharp as a blade against the sunrise. Above it, two tiny mare's tails of cloud were picked out in pink; to the south-east, a week-old moon was leaning back below the stars. I came up over the brow of the hill and the ground was spread with close-cropped heather and bare patches of dried-out soil. Granite sand was gritty underfoot, glowing pale in the half-darkness. When I walked across the heather it crunched beneath my boots. I tried to tread more quietly, and slowed my pace. It was then that I saw I was among them – all the slight rises in the heather around me were cairns.

Shipman Head Downs is the largest cairn field in the south-west of Britain. A hundred and thirty-four little monuments spread across this hill-top. Some look like clearance cairns and you can trace their lines through the heather. Others were for burial. Wandering among them, so many in the half-darkness, I felt I was crossing a battlefield, carefully placing each footfall so as to avoid stepping on the dead. At the further edge, the slope dropped away, down to the rocks of Hell Bay, and beyond that were the rocks and islets, no more than black spaces in the dawn-pale of the sea.

Whoever buried their dead here, they had the mariner's awareness of the shape of the coast. Some may have carried the bodies across from other corners of Scilly or by boat from the off-islands. Others might have ferried them out from the mainland of Cornwall. They brought them here in part because of those who had come before and left the dead and marked the place with stones. But each in their way came here too because of where it was – out on the far edge of the main Scilly Isle, out from Cornwall and Britain and the landmass of Europe, out here where the sun sets, the western-most place where land was no more than a whisper in the silence of the ocean, a flat-topped hill already half-way to eternity.

# EPILOGUE

Back home, the summer was in full swing. The lanes were full of cars, the beaches were full of people, and the answerphone was full of messages. My family were not due back from up east for another day or two and I busied myself around the place with catching-up tasks. I then rowed the boat up one of the side creeks to 'Crystal Pool' (so named by Clio) and pitched my tent on a grassy shelf of ground. In the twilight, the water reflected a sheet of feathery clouds and the first stars. At dawn, I woke to a sound like rain falling in the oaks, but it was dew, so thick it was now running down through the leaves. I lit a fire and dragged an oak bough across the shingle to sit on. The fire crackled, the kettle boiled and I sat drinking coffee and reading a book that I hadn't opened in nearly thirty years. I'd dug it out of my shelves by chance the day before. It was *Peter Camenzind* by Hermann Hesse. I'd loved it then, and was filled again with its naive and simple joy: 'At that time in my life, I did not know the names of the lake, mountains and streams of my native place. But I saw the blue-green water sparkling with tiny lights in the sunshine and, in a close girdle around it, the steep mountains…'

One of the messages on the answerphone had been from my friend Derek, a retired marine scientist. He was coming up that afternoon to do a survey of the bass nursery and could I help? I'd done a couple of surveys already with Derek. He had a twenty-nine metre beach-seine net and would come up in his boat with two or three others and we'd all wade the net up a gully, flood tide gushing around our thighs, and loop one end round to the bank again. We'd then kneel in the mud, pick out the 'year one' bass, count and measure them. It was a messy business, fingering the ooze for all those tiny creatures. You could tell the bass from young mullet or gobies or sandsmelt because they had a rainbow sheen to them and a dorsal line – like a mature bass in miniature.

The life cycle of bass is not unlike that of salmon, in that they each spend most of their lives in the open sea but have an ability, perhaps through smell, to find estuaries and swim far up along their tidal branches. Unlike salmon, they are born not in the rivers but offshore, where they drift around as plankton until big enough – about two centimetres long – to find the openings in the coast and swim upstream. They then spend some four years maturing. I found this homing instinct wondrous in many ways, not least because it brought them here, to the same creek where I'd ended up.

Later that day, I was rowing to our rendezvous when Derek phoned to say he had a snag with his engine. If he couldn't sort it in half an hour, we'd miss the tide and have to call it off. I carried on to the bend in the river, and dropped anchor. There was little wind. The stern soon swung round to lie with the tide. The bows settled, facing Penkevil Point and I took the opportunity to rethread the mainsheet and tend to a problem with the jib halyard. High above the mast, the woods rose from the water, and I looked up at the tiers of green – beech upon oak upon oak, with darker holly underneath, and a single sweet chestnut. Out of their canopies dropped the woodland sound of finches. A faint breeze hissed in their leaves. The sun bounced off the white deck and it was hot and the afternoon drifted by…

Then suddenly, piercing the clotted air, a whimbrel. Seven notes. That electric sound, like an alternating current coursing through the valley. It took me a moment before I spotted it, speckled brown and flying fast against a backdrop of leafy green. I watched it until it was gone, out of sight behind the point.

The creek reverted to its stillness. Opposite the wood, on the other bank, was the rise of Ardevora Veor, (*veor* meaning 'great'). Almost an island, its bulk pushed the Fal round behind it; on the opposite side was a low isthmus. The longer we spent here the more I found myself aware of that small hill – glancing at its distinctive slopes, its patches of woodland and its rounded-off top. I thought of what Vincent Scully said about the siting of all Cretan palaces – the 'gently mounded or conical hill on axis with the palace to north or south', and the tumulus in our field, south of the hill – and it made me smile at myself. Often, here, I had the sensation of being right at the centre of the land. I remembered something similar at other places, at certain spots in our field, at Leskernick on Bodmin Moor ('the *omphalos* of the saucer'), and on Scilly, the westward view from St Martin's of the flooded interior of the old island.

It can have several effects, that impression of being enclosed by the rising ground. It can help populate its features with ancestral spirits or deities. It can lead those of a certain disposition into far-reaching theories. It can swell your own presence to such a size that you become the sole agent of all that surrounds you, a giant of solipsistic intensity. Or, as on that still and sleepy afternoon, you can feel the self dissolve, let its burden drop, and slip back into the slopes and hollows of the land.

Silence again. Then, from far off down-river, I heard the faint sound of an outboard, and within minutes Derek's Orkney appeared between the wooded banks. I saw his hand waving above the cuddy. Stepping up into the bows, I started to haul up the anchor.

# NOTES

For the toponymic notes at the head of each chapter, I am indebted to the incomparable work of Oliver Padel (Padel, 1985, 1988), President of the English Place-name Society, with a particular interest in Cornwall. Craig Weatherhill's work (Weatherhill, 1998) has also been a valuable resource. It's a habit that's hard to stop once started – breaking up place names and trying to work out what they might mean. In Cornwall, it has a double appeal – offering glimpses of long-ago perceptions of places and of the old language itself.

## 1. MENDIP

The article about Aveline's Hole, 'UK's Oldest Cemetery Identified', by Maev Kennedy, appeared in the *Guardian*, 24 September 2003.

The very thorough analysis of the remains found at Aveline's Hole is in Schulting (2005) pp. 171–265; its title, '... Pursuing a Rabbit down a Hole', refers to the original discovery of the cave: in 1797 two young men were chasing a rabbit in Burrington Combe and when it disappeared down a hole, they dug down and discovered a cavern filled with propped-up skeletons. Boycott and Wilson, 'In Further Pursuit of Rabbits: Accounts of Aveline's Hole, 1799 to 1921' (2011) pp. 187–233, contains reproductions from the journals of Reverend John Skinner; the sketch of the interior of the cave is on p. 207. For *siejddes*, see Bradley (2000) pp. 6–7.

## 2. ARDEVORA

Martin Heidegger's essay 'Building Dwelling Thinking' is published in his collection of essays, Heidegger (1975). The quotes about the idea of place come from Escobar (2001) p. 143, from the Preface to Edward Relph (1976), and from Edward Casey (1996) p. 18. W. H. Auden's neologism 'topophilia' appears in his introductory essay to John Betjeman's *Slick but not Streamlined*, Auden (1947). Tim Cresswell's *Place: A Short Introduction* (dedicated to Yi-fu Tuan) is a good first stop for looking into the question of place; the quote is from Cresswell (2004) p. 23. The landscape preferences of Yi-fu Tuan are from the Preface to the Morningside Edition, Tuan (1990) p. ix, as is the 'Without exception, humans grow attached to their native places' (p. xii). The 'vertical' and 'horizontal' perceptions are presented on pp. 129–49.

Wordsworth's 'stepping westward seemed to be / a kind of "heavenly" destiny' is from 'Stepping Westward' in *Memorials of a Tour in Scotland, 1803*, see Wordsworth (1895) p. 289.

Walter de la Mare's unease in Cornwall is cited in Val Baker (1973) p. 8. The quote from Hammond Innes is from *Wreckers Must Breathe* (1959) p. 123. Ruth Manning-Sanders's 'sense of the primordial...' is quoted in Val Baker (1982) p. 12. Jacquetta Hawkes wrote the words about Cornwall in the form of a poem recited in *Figures in a Landscape* (1953), a film about Barbara Hepworth's work. The scene from Daphne du Maurier's childhood comes from *Vanishing Cornwall* (Du Maurier, 1967) p. 3. Herring's thesis is *An Exercise in Landscape History: Pre-Norman and Mediaeval Brown Willy and Bodmin Moor, Cornwall* (Herring, 1986).

## 3. BODMIN MOOR

For a full list of Dorothy Dudley's archaeological work, including her work on Garrow, see www.biab.ac.uk/people/27474. Her archive is in the Courtney Library, Truro. Cranborne Chase is discussed in Tilley (1994) pp. 143–201, Rudston's ritual landscape is in Clark (2004). The spread of sacred sites on the mountains of Attica is from Pausanias (1898), Book 1.32.2. The quotes on Hindu sacred landscape are from Eck (2012) pp. 4–5. For Stowe's Pound see Tilley (1996) pp. 161–176 and Bradley (1998) pp. 13–22, and also Davis (undated).

For W. G. Maton's thoughts on the Cheesewring, see Chope (1967) p. 268. George Borrow on the Cheesewring is quoted in Walling (1909) p. 159. Borrow was invited by his Cornish cousins to visit them just to the south of Minions Moor. From there he conducted a typically energetic walking tour of Cornwall. He claimed some years later that he had a two-volume book on Cornwall 'ready for the press'. Some of the Cornish notebooks survive (in the library of the Hispanic Society of America in New York) and have been abridged and published in a limited edition of thirty copies by the George Borrow Society, Fraser (1997), along with relevant correspondence. For Wilkie Collins at the Cheesewring, see Collins (1851) pp. 42–3.

For Daniel Gumb, see Paynter (1931–6), vol. IV, issue II, pp. 1–4. In the *Journal of the Royal Institution of Cornwall* (1873), pp. xx–xxi, is a note on the destruction of Gumb's original house which had occurred since the institution's excursion there in 1868.

## 4. GARROW TOR

The Leskernick project was written up extensively in *Stone Worlds*, Bender et al (2007). The Jan Farquharson poem is on pp. 277–9 and also includes the lines: 'The Bronze Age really changed how this hill went. / It seems they left no weapons, beads or pots / but that's alright, the stones were what they meant.' The quote from Alfred Watkins comes from the Preface of the original edition of *The Old Straight Track*, Watkins (1925) p. v, while Vincent Scully's discussion of the siting patterns of Cretan palaces is in Scully (1962) p. 11. Much of Roger Farnworth's work on Bodmin Moor is contained in his long article 'The Focus on Rough Tor and Stowe's Pound in the Neolithic and Early Bronze Age' (with Peter Herring and Bryn Tapper) – to be published.

## 5. ROUGH TOR

For Cornish place names, I have used Padel (1985, 1988) and Weatherhill (1998). The debate about Bun na hAbhann in *Translations* (Friel, 1982) is from Act 2, Scene 1, pp. 34–5.

For sacred mountains, see Tuan (1990) pp. 70–74, Eliade (1963) pp. 99–102. The Zoroastrian quote is from *The Zend Avesta, Part III* (Mills, 1887), *Yasna* 6.13, and from Herodotus (1954), Book I, p. 131. For Mandelstam's 'Ararat' sense, see Mandelstam (1980) p. 57. For the indigenous beliefs of Mongolia, see Heissig (1980). My own journey to Otgen Tenger is told in *Granta 83: This Overheating World*. Two months after our journey, not far from Ulan Bator, Narmandakh and his father were killed in a car accident. Malinowski's definition of myth is from Malinowski (1992) p. 87.

For Sabine Baring-Gould's assessment of the bogs below Rough Tor and Brown Willy and his wish to see commercial 'promoters... flounder there', see Baring-Gould (1981) p. 124. For Rough Tor as 'fortress' and 'defended enclosure' see Hencken (1932) p. 100, and Johnson and Rose (1994) p. 46. Rough Tor's bank cairn – and the significance of Rough Tor itself – is discussed in Pryor (2010) pp. 68–74. Pryor was part of the Channel 4 Time Team who dug a section of the Bank Cairn; see Thorpe (2011) for an outline of the Time Team work.

## 6. TINTAGEL

The land charter for Tintagel's sale to Richard, Earl of Cornwall in May 1233 is at the Public Record Office in Kew (PRO E36/57, folio 44v, no. 163). A very full discussion of Tintagel's early chronology, and the archaeology of the Island, the castle and Tintagel parish churchyard can be found in *Cornish Studies 16*, Thomas (1989). For biographical details of Richard, Earl of Cornwall, see Vincent (2008). The story of Arthur's conception is in Geoffrey of Monmouth (1958), Book VIII, 19, pp. 174–7. The visit to Bodmin of the Canons of Laon in 1113, and their insensitive suggestion that Arthur might not still be living is told in a manuscript of Hermann of Tournai, see Padel (1994) p. 9. The pottery finds at Tintagel as evidence of patterns of post-Roman trade to Britain are discussed in Thomas (1988) pp. 7–23. For an understanding of Tintagel, I am also grateful for private conversations with Charles Thomas and Carl Thorpe.

Reference to early ownership of the manor of Bossiney can be found at the Public Record Office in Kew (Pipe Roll 10 Richard, p. 175, 1198; Pipe Roll 9 John, p. 77); see Padel's chapter, 'Tintagel in the Twelfth and Thirteenth Centuries', in Thomas (1989) pp. 61–6 for a full list of references. For the range of Arthur place names in Britain, see Green (1999) pp. 1–19, while Padel's analysis of the Arthurian cycles and his suggestion of Arthur's origins as a 'pan-Brittonic figure of local wonderment' is in Padel (1994) pp. 1–31.

The description of Tintagel's unassailable position, the tenuous attachment of the island upon which its mythology is based, is in Geoffrey of Monmouth (1958), Book VIII, 19, p. 175. For the expansive capacity of islands – 'the sea elevates these few acres into something they would never be if hidden in the mass of the mainland' – see Nicolson (2001), p. 141.

The quote on the importance of Geoffrey's *Historia Regum Britanniae* (c.1136) in the Middle Ages is from Kendrick (1950) p. 7; for Elizabethan plays based on the British History, see ibid. p. 39. Geoffrey of Monmouth's idyll: 'Forests also hath she filled with every manner of wild animal, in the glades whereof...' is in Geoffrey of Monmouth (1958) Book I, p. 4. For Milton's reference to Geoffrey's myth of origin, see the extract from *The History of Britain* 1670, ll. 15–17, in Milton (1900); some sources say that Brutus was not Aeneas's grandson but his great-grandson.

For Gildas's 'brilliant rivers that glide with gentle murmur, guaranteeing sweet sleep for those who lie upon their banks' and his vision of Britain as a *locus amoenus* see Gildas (1978) pp. 16–17. For Bede's similar version, see Bede (1969) Book I, pp. 14–17.

The drily prescriptive definition of the *locus amoenus* is from Curtius (1953) p. 195. See Clarke (2006) for an intriguing study of the *locus amoenus* and the medieval conflation of idealised topography and national identity. The quote from Shakespeare's *Richard II* is from Act II, Scene 1. 'Dismal is this life...' is from Jackson (1935) p. 13; 'The harp of the wood plays melody...', ibid. p. 23.

## 7. GLASTONBURY

St Dunstan's dream is told and interpreted in the fourteenth-century manuscript *The Chronicle of Glastonbury Abbey* by John of Glastonbury, see Carley (1985a) p. 128. The dream originally appears in Wulfstan of Winchester's *Vita Aetholwoldi*, then was retold in William of Malmesbury's twelfth-century *Vita Dunstani*. See Clarke (2006) pp. 67–78 for a full discussion of Glastonbury's miraculous topography and its representation as a 'mirror for England' as a whole. The suggestion of Glastonbury itself as a *locus amoenus* is from Stubbs (1874) and also quoted on an interpretive board at the ruins of Glastonbury Abbey. St Collen's oft-told spirit-journey comes from the sixteenth-century Welsh text *Buchedd Collen*, Hafod MS 19 (1536), Cardiff Central Library.

The strange story behind the seance and the rediscovery of Glastonbury's Edgar Chapel is told in Bond (1918). For Glastonbury's zodiac, see

Maltwood (1934). James Carley (1988) pp. viii–ix tells the story of cataloguing the Glastonbury Zodiac papers in 1969, contained in their 'black velvet case' which was heavily sealed in the Maltwood Museum, Victoria, British Columbia, Canada. From there he flew to the UK and reached Glastonbury for the first time, where 'an ancient woman threw her arms around me and greeted me as Bors', Carley (1988) p. ix. It is pleasing somehow that the career of Professor Carley, one of the most rigorous scholars of medieval Glastonbury, should originate in a show of its less rational aspects.

John Cowper Powys's panegyric from *A Glastonbury Romance* is quoted in Carley (1988) p. viii. For the Steinbeck letters, see Steinbeck (2001) January, February 1959.

For Leland's verses at the royal wedding of May 1533, see Toulmin Smith (1907) p. ix. Leland's own account of his first sight of the library of Glastonbury Abbey is in Carley (1985b) p. 142. Leland's promise to the King of how he would explore the kingdom, missing neither 'cape, nor bay, haven, creke or peere…' is published in Toulmin Smith (1907) p. xli as part of 'The Laboriouse Journey and Serche of Johan Leylande for Englandes Antiquities, Geven of Hym as a Newe Yeares Gyfte to Henry the viii in the xxxvii yeare of his raygne'. This title was added by Leland's editor, John Bale, in 1549. The document itself in Leland's own hand is untitled (Leland's *Collectanea* vol. III, Bodleian Top. Gen. C3, p. 281) but is generally known as Leland's 'New Year's Gift'.

William Harrison's complaint that Leland's folios were 'moth-eaten, mouldy and rotten…' is quoted in Kendrick (1950) p. 149ff. For Leland in Paris, see Carley (1986) p. 19. John Dee's lament for Glastonbury is found in his *General and Rare Memorials Pertaining to the Perfect Arte of Navigation* (1577) quoted in Carley (1988) pp. 169–70. Michael Drayton's topographical epic *The Poly-Olbion: A Chorographical Description of Great Britain* (1622) is quoted in ibid. p. 170. Leland's rapturous response to being shown the relics of Arthur and the lead cross from Arthur's original grave is recounted in Kendrick (1950) pp. 96–8.

For prehistoric Glastonbury, see Rahtz (1991) pp. 3–37. The full quote from William of Malmesbury is: 'In the meantime it is clear, that the depository of so many saints may be deservedly styled an heavenly sanctuary upon earth. There are numbers of documents, though I abstain from mentioning them for fear of causing weariness . . .', Giles (1847) p. 23. Leland's plans to present the King with dozens of volumes of works are listed in his 'New Year's Gift', see Toulmin Smith (1907) pp. xxxviii–xliii. The theories about Leland's mental illness can be found in Carley (2006) p. 4.

## 8. HENSBARROW

For the range of Cornwall's minerals, see Payton (2004) p. 10: 'the 440 species of minerals in Cornwall so far represent a startling 15 per cent of the entire world's total'. For the quality of the china clay around St Austell being 'the finest in the world', see Barton (1966) p. 20.

Charles Causley's description of the clay country as a 'weird, white world dusted over with the colour of sex' comes from his Introduction to Jack Clemo's *The Map of Clay* (Clemo, 1961) p. 7.

Details of tin ventures in what was to become the clay country can be found in the Cornwall Record Office, CRO J/1/159 1752–3, and regarding Cookworthy, offering him 'liberty, license and authority to search . . .', CRO, Fortescue Records, Acc. No. 342 1770, referenced in Barton (1966) p. 21. The description of Cookworthy as 'one of the greatest chymists this country has ever produced' appeared in an obituary in Felix Farley's *Bristol Journal*, 21 October 1780, see Selleck (1978) p. 20. For general material on Cookworthy's life, see Harrison (1854), Penderill-Church (1972), Selleck (1978), and Winchester (2004).

For Neolithic axes and the quarry at Pike O'Stickle in Cumbria, see Bradley (2000) pp. 85–7. Janet Gleeson's *The Arcanum* (1998) gives a full account of Böttger's efforts to produce porcelain, as well as details of the Europeans' high regard for it. Père d'Entrecolles' two letters can be found in Burton (1906), chapter IX.

Cookworthy's letters to Pitt are in the Cornwall Record Office, Fortescue Papers DDF (4) 80/21; they also contain the samples of his endeavours to create porcelain, as well as many details of experiments with the design of chimneys and different fuel. For the general history of the china-clay industry, see Barton (1966) and Hudson (undated). During this period of experimentation, other sources of china clay and china stone were available. Cookworthy had also discovered clay and moorstone from Tregonning Hill near Helston and some material was also arriving from Virginia. But one geological historian has stressed: 'In the light of our present knowledge of the materials, it may be suggested that the secret of Cookworthy's success lay in his use of the excellent china stone from the Carloggas district of St Stephen', Pounds (1948), pp. 20–33. Samples of William Cookworthy's porcelain are on display at Plymouth City Museum; see also Mackenna (1946) for illustrations of his potteries' work.

Cookworthy's own exposition of his divining knowledge is in Pryce (1972) p. 116. The two accounts of Cookworthy's divining experiment laid on for Joshua Reynolds and Samuel Johnson are told in Selleck (1978) pp. 44–5. The Swedenborg quote about forming man out of 'the dust from the ground' is from Swedenborg vol. I (2009) p. 60. 'I have the flesh, the bones and the sinews but to put flesh on bone to make a whole body, that is what perplexes me' is from a letter to Dr Richard Hingston of Penryn; see Penderill-Church (1972) p. 45.

## 9. FAL

'A landscape of purgation' is from Clemo's *Confession of a Rebel* (1949) p. 5. 'The clay landscape dominated me from childhood…', letter from Jack Clemo to Alan Kent, 21 October 1987, displayed at Trethosa Chapel. The full quote of Charles Causley's assessment of Jack Clemo is: 'Jack Clemo once described himself as the oddest writer Cornwall has produced. I would describe him as one of the greatest.' Reading his poems, Causley 'never doubted' that he was 'in the presence of a man whose make-up includes genius'. See Clemo (1961) p. 11.

'The soil was thrown into tanks and kilns…' Clemo (1949) p. 6. 'There were no rhythms about it, no recurrences; only a pitiless finality in every change,' ibid. p. 6. The 'fantastic swellings and angles of the clay landscape', ibid. p. 208; the phrase comes from Clemo's notion of 'a sense of spiritual congruity' between the clay country and 'the savage distortions' of the paintings of Picasso – he had bought a book of Picasso with early proceeds from his writing. Clemo's dismissal of 'nature worship' and the Romantics is in ibid. p. 200. '*His* hand did not fashion / The vistas these poets admire…' is from 'Neutral Ground', Clemo (1961) p. 20. Those who glimpsed the eternal in the earth's beauty displayed for Clemo 'a slackness of fibre', Clemo (1949) p. 200.

'It became a familiar and pathetic sight to the people of St Dennis, the solitary widow, dark, tight-lipped...' Clemo (1949) p. 26. Clemo's bout of blindness in his early teens, in which he sat on a board suspended above the stairwell – 'like a captive bird' – is described in Clemo (1949) pp. 58–63.

'The Clay Verge' is in Clemo (1951) and 'The Map of Clay' is in Clemo (1961); *The Map of Clay* was published by Methuen and contains within it the poems from *The Clay Verge* (Chatto & Windus). 'The Two Beds' is in Clemo (1961) pp. 47–8. 'I am beyond your seasons: food / For these is in your blood but not in me...' is from 'Cornish Anchorite' in Clemo (1961) p. 41.

Clemo's dreaming of 'strange gods' on Bloomdale clay-tip is from Clemo (1949) p. 18. His grappling with 'the tide of prayer... and the black tide of lust' is from ibid. pp. 155–6. His wartime wanderings and the 'lonely vigil' he kept on the clay-tips while 'the tears would stream down my cheeks as I moved through the rain-soaked bracken or up the wet sandy slopes' is from ibid. pp. 183–4. The lines from 'Beyond Trethosa Chapel' are in Clemo (1961) pp. 63–4. 'The mystical-erotic quest' was a phrase he used in the subtitle of a later autobiographical work, *The Marriage of a Rebel* (Clemo, 1980), in which he also sketches out how he and Ruth Peatty met (ibid. p. 128). *The Invading Gospel* (1958), which he sent to Ruth by way of introduction is a bracing account of his own odyssey of faith. Clemo's summing-up of his own redemption – 'I lived in the summer sunshine of fulfilled hopes...' – appears as a note to the third edition of *Confession of a Rebel* (1988) p. x, adding: 'I almost ceased to recognise the suffering misfit whose story is told in these pages.' The final quote from 'The Clay-Tip Worker' is in Clemo (1951) p. 33. Trethosa Chapel closed in 2013 and the replica of Clemo's cottage, as well as other effects, were transferred to Wheal Mertyn China Clay Museum.

## 10. TREGONY

Clemo explains how he responds to others' worship of spring: 'I advance to pour / Sand, mud and rock upon the store / Of springtime loveliness idolaters adore' – from 'The Clay-Tip Worker' in Clemo (1961) p. 37.

It was W. Penaluna (1838, vol. 1, p. 141) who noted that the Fal 'excels all the harbours of the isle except Milford Haven.' Charles Henderson suggests that 'Tregony in the 12th and 13th centuries was perhaps the chief sea port in Cornwall' in Henderson *Ecclesiastical Antiquities* vol. V, 'Powder', p. 436. For the castles at Trelonk, and the oral tradition of one at Ardevora, see Henderson *Ecclesiastical Antiquities* vol. V, Ruan Lanihorne and Philleigh parishes. The Henderson Collection, both the manuscripts he collected and the extensive notes he made for his own unpublished works, are kept at the Courtney Library, Truro (see Chapter 12 below).

A good overview of the Fal Estuary's history, its silting and its shoreside buildings can be found in Ratcliffe (1997); see also 'Some Creeks of the Fal', *Falmouth Packet*, 6 September 1929. Whitley (1881) vol. VII, part 1, p. 13 notes that opposite Ardevora, in 1698, there was six feet of water at low water springs. For Tregony's early importance: 'Tregony, no doubt, was the very first town on that magnificent haven now called Falmouth Harbour, the Cenionis Ostium of Ptolemy', see Polsue (1974) p. 282.

The will of Robert Bennett is in the Cornwall Record Office, CRO AP/ B/155/1–7. Christine North has written a piece about Bennett's will and others in *Journal of the Royal Institution of Cornwall*, North (1995), pp. 32–77. Richard Mabey's journey to the upper Fal Estuary is recounted in Mabey (1999) pp. 31–4.

Whitaker's intention not to add to the 'private and dull annals' is from Whitaker (1773) vol. 1, p. ix. For general biographical information on John Whitaker, see Sweet (2004) and Polwhele (1831) vol. III. For a physical sense of Whitaker – including his ebony teeth – see ibid. pp. 69–70. The Tolstoyan picture of Whitaker reaping with his parishioners: 'in the fields from 9 to 7…' is from a letter dated 22 August 1796, transcribed in ibid. p. 178. The 'fifty fair maids', details of marriage and family are from ibid. p. 70; Whitaker being good with children, ibid. p. 164. The list of publications he wrote for following his exile to Cornwall is from Sweet (2004) p. 3.

Whitaker's manuscript, 'History of Ruan Lanihorne', is in the Courtney Library, Truro (Whitaker MS A1). H. L. Douch transcribed much of it and published it in the *Journal of the Royal Institution of Cornwall*, Whitaker (1974). In the final paragraph, Whitaker makes the claim for his work's originality, 'I have thus with a hasty hand sketched out a draught of a Parochial History in a single parish… It is the first draught I ever saw of the kind', in ibid. p. 152. In the same paragraph, he ends with a florid signing-off, as if from his own pulpit he is proclaiming an antiquarian's manifesto. A place like Ruan Lanihorne appeals to those 'who therefore love to contemplate the records of its former fortunes by the shadowy light which antiquarianism can throw over them, and enjoy an intellectual memory in walking amid the monuments of other days by the half-revealing beams of this historical moon. Jan 17.1791' (ibid. p. 152). In the manuscript, the date is written over the top of a crossed out 'December 4 1790' – so his hand was not quite so 'hasty' as he suggests. Dr Johnson's quote on Whitaker is from Douch's introductory note to his transcription, Whitaker (1974) p. 108. Details of the village and the castle, and Whitaker's toponymy, are from the pages of the transcription, ibid. The Archdeacon family were also known as Lercedekne; for the 1334 licence to 'crenellate' their house see Turner and Parker (1859) vol. 3, part 2, p. 410.

The picturesque movement is given lively and detailed analysis in Christopher Hussey's work on the subject, Hussey (1967), in which William Gilpin's books, and William Combe's satirising of them, are discussed, pp. 111–24. The elegiac passage from Uvedale Price is quoted in Piggott (1976) p. 120. Gilpin's reluctance to explore 'uninteresting' Cornwall is told in Gilpin (1808) pp. 190–97.

Details of Whitaker's last trip to London are in Polwhele (1831) vol. III. Richard Polwhele's praise for his friend Whitaker pours from the pages of this volume, which is given over completely to Whitaker's life and a selection of his letters. Whitaker's 'overpowering' intellect is from ibid. p. 3, where Polwhele also notes Whitaker's pre-eminence: 'we can get no Cornish worthy upon a level with Whitaker in conversing, in writing, in acting'. The comparison of Whitaker's sentiment to a craggy upland scene is in ibid. p. 161.

For Ogilby's highway map of 1675, see Ogilby (2005), Sheet 4. Robert Morden's map of 1701, reproduced in Cox (1720), shows this route – through Tregony and across King Harry Passage – as being the main road in and out of Cornwall, possibly a reflection of Falmouth's rise in importance once it became the country's principal packet port in 1688.

## 11. RUAN CREEK

For Thoreau's pleasure on looking at his own wood-pile see Thoreau (1986) p. 296. Aldo Leopold's sawing of the oak comes from Leopold (1949) pp. 6–18. For *A Street through Time,* see Millard and Noon (1998).

The Fal Valley railway project map can be seen in the Cornwall Record Office, Truro, CRO QSPDR/5/10, OS 1-inch map.

For barrow digging in the eighteenth and nineteenth centuries, see Marsden (1974): Fausset pp. 6–9, Cunnington and Hoare pp. 11–26. Skinner's poem was fully titled 'Lines written on the escape of a bee from his imprisonment of 2000 years, to a fair lady who was the cause of this emancipation. January 24th 1826', ibid. pp. 26–7. The lines from the Reverend Charles Woolls's 'Song of the First Barrow Digger' are quoted in ibid. p. 29. For barrow digging, see also Trigger (1989) pp. 66–7.

For the original discovery of the gold cup in Rillaton barrow, see Smith (1936) pp. 1–3. The cup is of ribbed gold and is thought to be of Mycenaean origin. An exact copy of it is kept at the Royal Cornwall Museum, Truro.

For biographical information on Baring-Gould, I have used Colloms (2005), Dickinson (1970) and Purcell (1957), also Baring-Gould (1923). Baring-Gould's 'love' and 'obsession' for the moors is in Todd (1987) p. 24. Charles Causley's assessment of Baring-Gould's 'feeling for the particular ambience of a place…' comes from his Introduction to Baring-Gould's *Cornwall* (1981) p. iii. For 'seity' see ibid. p. 242. Details of Baring-Gould's find at Pau, and sketches of the mosaics and motifs can be found in an 'intriguing notebook' that came to light in 2005 at the back of a bookcase in the home of Baring-Gould's great-granddaughter, see Wawman (2010) pp. 2 and 28–35. The two incidents from Baring-Gould's early life around Pau in 1850 are told in his autobiography, Baring-Gould (1923) pp. 172–3.

For late silting of the Fal, see CRO W. H./374, and the 1813 ordnance survey (sheet 31) CRO AD/819/3, in which the tidal part of the Fal still extends, like an engorged leech, up towards Tregony.

## 12. TOLVERNE

For Sir John Arundel of Tolverne, see Johnson (1958) pp. 141–5.

The story of Charles Henderson and his own work can be found in the hundred-odd boxes of the Henderson Collection, Courtney Library, Truro. The collection is an astonishing testament to his energies and includes the manuscripts he gathered from around Cornwall, his personal papers, his articles and his own notes, divided into *Antiquities* (roughly pre-Christian) and *Ecclesiastical Antiquities* (roughly post-Christian); in its handwritten form, this remains one of the most detailed accounts of the county ever produced. Charles Henderson's proof of Henry VI's association with King Harry Passage and the chapel in the woods below Tolverne is in the Courtney Library, Truro, Henderson *Ecclesiastical Antiquities* vol. V, Philleigh parish, p. 405. Additional material, including the letter from Cardinal Gasquet, is in Henderson *Ecclesiastical Antiquities*, vol. II, p. 43. A record of the miracles attributed to Henry VI survives in a manuscript in the British Library (Royal 13c. VIII) and a translation of a number of them was published by Knox and Leslie (1923).

For a physical description of Henderson, see Rowse (1986) p. 291. It comes from Rowse's address at the unveiling of a monument to Henderson in Truro Cathedral on 25 October 1983, fifty years after Henderson's death (ibid. pp. 291–6). Henderson's interest in postmarks, 'At a very early age…', comes from Henderson's own 'Apologia pro Labore sua' (hereafter Ap. Lab.), an Introduction he wrote to his 'Key and Index to my Documents', dated 24 September 1924 – although the Ap. Lab. was written, by his own account, two years earlier on 9 September 1922: Henderson Collection, Courtney Library, quoted in Henderson (1935) pp. xvi–xvii.

The 'snapshots of friends, boats, picnic parties…', Ap. Lab. pp. 3–4; 'obsessed with the idea that the time had come to sweep old corners clean…', ibid. pp. 4–5; 'In an unoriginal time when everything seems to have been tried before, this was a delightful sensation', ibid. p. 5. Henderson ended his apologia with the prescient words: 'Modern invention is still powerless to regulate our entry into or departure out of this world. Should I be prevented from continuing my work I leave these directions to those that come after,' ibid. p. 9; 'I ransacked the Public Record Office…' ibid. p. 5.

'It was clear that he had been everywhere…' is from the *Pelican Record* (Corpus Christi College record) vol. XXI, no. 4, December 1933. For the story of Sir Robert Edgcumbe going up to New College, see Henderson (1935) p. xviii, and for 'Henderson has not seen it yet'.

'He had a wonderful sense of topography…' is from Rowse (1986) p. 292. Rowse as 'the first working-class man…' is from Ollard (1999), cover copy; 'dear Charlie' is from Rowse (2003) p. 100; 'He certainly had plenty of fun…' ibid. p. 408; 'a gangling great boy…' ibid. p. 408. 'There are fifteen places called Penquite east of Par…' Rowse (1986) p. 292. Rowse's challenge to Henderson to identify the parish of any given Cornish farm is from an obituary by Rowse in the *West Briton*, 5 October 1933. 'I had to take his *general* reading in hand . . .' Rowse (2003) p. 408. Rowse's assessment of Henderson's sexuality is in ibid. p. 409. 'He was developing, flowering, restless, catching up on life . . .' from Rowse (1986) p. 295. 'On his deathbed he looked like an Angevin king', quoted from a letter to A. L. Rowse from Henderson's widow Isobel, Rowse (2003) p. 100.

The Spoffkin material is found in various school exercise books, sketch books and a gold-embossed green notebook entitled *Spoffkin Graphic, 1912* in the two boxes of personal papers in the Henderson Collection. Baring-Gould's *The Curious Adventures of Dr Roticher* is in Wawman (2010). The imaginary places of the young Brontës are presented and discussed in Brontës (2010). Rowse's final thought on Henderson the topophile: 'What led him on was that he loved every hill and valley. . . .' is from his obituary in *West Briton* 5 October 1933.

For thoughts on the early name of King Harry Passage, see Henderson *Articles*, no. 6, King Harry Passage. For *tirthas*, see Eck (2012) pp. 7–12, also pp. 452–3: 'the divine, though utterly transcendent and ultimately ungraspable by human mind and speech, takes form in the very world in which we live – in rocks and hillocks, in rivers and pools . . . God is vast, yet God is here.'

## 13. PORTHLEVEN

For Eliot's assessment of the New England mountains as '. . . more desperate than the desert' see Eliot (1934) p. 17. Naum Gabo saw paintings 'in a torn piece of cloud carried away by the wind . . .': letter to Herbert Read, 24 June 1958. Ben Nicholson's statement, 'Can you imagine the excitement which a line gives you when you draw it across the surface . . .' is quoted in Val Baker (1973) p. 18.

A view of the Cornish roots of his painting was given by Peter Lanyon's cousin, Rosalie Mander: 'Here is the work of a true Cornishman, born & bred in West Penwith, not one of the cuckoo orphans come down to claim the home . . . Peter Lanyon's work has a backbone of granite underneath its charm; when this trips up the foreigner there is a chuckle of laughter on the Downs . . . You take risks here, Stranger', from the Foreword to the catalogue of a 1951 exhibition in St Ives: *Paintings from Penwith by Peter Lanyon*. 'How they natter at my feet…', letter to Ivon Hitchens, probably late 1952, Lanyon (1990) p. 133.

Lanyon was described as 'the last landscape painter' by the critic Lawrence Alloway, quoted in Causey (2006) p. 190. Lanyon as a self-confessed 'place man' is quoted in Garlake (1998) p. 7, and Lanyon's own 'I paint places but always the Placeness of them' is from a letter to Paul Feiler, c.1952, in Lanyon (1990) p. 125. 'I'm really just an old landscape painter like Constable…' is from a recollection by Michael Canney, ibid. p. 160. For extensive material from Peter Lanyon's letters and notes, as well as from interviews with him and reflections by critics and fellow artists, Andrew Lanyon's albums are an encyclopaedic and evocative source, Lanyon (1990, 1995, 1996, 2006a); all these works are privately published, see www.samlanyon.com/andrewlanyon/booklist.htm.

For Lanyon playing dominoes in St Ives pubs, see Stephens (2000) p. 13. Peter Lanyon's skit about Ben Nicholson, 'TIN PAN BEN with a squeaky scream…', comes from the letter to Ivon Hitchens above, Lanyon (1990) p. 133. At the time of the letter, Lanyon was clearly distressed – frustrated with his work, under financial pressure, with three young children, and scarred by the split with Nicholson and others of the St Ives artists: 'I am BROKEN ARTED…I get so depressed sometimes I feel like murder,' Lanyon (1990) p. 133. Pissing against the wall of Nicholson's house, Stephens (2000) p. 86. 'Why don't you admit you're an abstract painter, instead of all this stuff about Cornwall?' is from Stephens (2000) p. 89. For the phrase 'a mile of history in a gesture', see Lanyon (1990) p. 216. Peter Lanyon equating his painting with the experience of rummaging in RAF dumps during his war-time service is from an undated note in Lanyon (2006a) p. 291: 'My painting has developed in precisely the same way…'. 'Old stone to new building…', lines 5–7 of 'East Coker', *Four Quartets* (1944), included in Eliot (1969) p. 177.

*The Yellow Runner* is discussed in Stephens (2000) pp. 47–60; see also Lanyon (1990) p. 72 for Peter Lanyon's own notes about it – 'the pollen and the flower, the sperm and the egg'. The wartime letter to Lanyon's sister – 'Do you know what that coast means to me? It means the sea…', is quoted in Stephens (2000) p. 47. For his need for speed and danger – 'Without the urgency…' – see Lanyon (1996) p. 18. 'The whole purpose of gliding,' he wrote, 'was to get a more complete knowledge of the landscape,' Stephens (2000) p. 155. Lawrence Alloway's enthusiastic response to Lanyon's 1960 exhibition at Gimpel Fils – 'the old works are like puddings' in comparison – was published as 'Light Waves' in *Weekly Post*, 29 October 1960.

The three quotes, 'I believe that landscape painting, the outside world of things and events larger than ourselves, is the proper place to find our deepest meanings', 'those places where our trial is with forces greater than ourselves...' and 'Landscape painting is not a provincial activity...' come from notes for a lecture Peter Lanyon gave for the British Council in Czechoslovakia, January 1964.

For Andrew Lanyon's satirical series of Walter Rowley books, see www.samlanyon.com/andrewlanyon/booklist.htm. Art as 'an infectious disease visually transmitted... a vivid fungus which only attacked mankind... art has repeatedly reached plague proportions, causing civilisation after civilisation to collapse', Lanyon (2006b) p. 23. Rowley's discovering that in St Ives 'there were so many artists working there that it became increasingly difficult to take a photograph without another photographer or a painter or a sculptor getting in the picture' is from ibid. p. 83. Alfred Wallis watching Nicholson and Wood 'like Montezuma observing the approach of the conquistadors' is from ibid. p. 89.

The painting *Porthleven Harbour*, and its troubled production, is discussed in Stephens (2000) pp. 72–80. Another record of Peter Lanyon's frustration with the picture comes from a letter he wrote at the time to Rosalie Mander, 6 January 1950: 'I lost everything about an hour ago and threw my oil stove at Porthleven... Luckily it wasn't alight.' Lanyon (1990) p. 100.

## 14. MORRAB

'It is not just a place, it is a *mysterious* place,' said Val Baker (1982) p. 12 of West Penwith, and went on: 'Strange, ghostly, haunted, solitary, menacing, brooding, awe-inspiring, weird, magical – the adjectives come tumbling out and yet somehow they are never enough.' Peter Lanyon's quote is from ibid. p. 144. Katherine Mansfield's view of Cornwall as 'not really a nice place...' comes from a letter she wrote to S. S. Koteliansky in May 1916.

John Heath-Stubbs, 'This is a hideous and wicked country...' is from 'To the Mermaid at Zennor' in *A Charm against the Toothache*, Heath-Stubbs (1954).

The toponymic roots of 'Cornwall' and 'Penwith' are discussed in Padel (1988) pp. 72–3 and p. 136. John Davidson's 'I felt the time had come to find a grave...' is from the 'Epilogue – The Last Journey', see Davidson (1908) p. 146.

Jeremy Le Grice's personal reflections on the St Ives artists – including Peter Lanyon – were given at a lecture, 'The First Eleven: St Ives Artists', at the Royal Institution of Cornwall, Truro, 11 October 2008.

William Borlase's archive in the Morrab Library consists of three letter books, with copies of letters he sent from 1722 to 1772, as well as six volumes of received letters, and a supplementary book of the same. For a full list of the Morrab's Borlase archive as well as other sources for William Borlase, see Pool's very good biography (1986) pp. 285–97. The letter on the joys of living in West Penwith, 'I have had the pleasure...', was written to George Borlase on 18 December 1727 (Morrab LB I 21). On his fifty-first birthday William Borlase wrote the letter 'I am preparing for old age, that is, laying in a fund of amusements...' to Dr Andrew, 2 February 1746/7 (Morrab LB II 12). Borlase's letter to his fellow Cornish parsons in 1752 is in an Appendix, Pool (1986) pp. 298–301.

For William Stukeley and the Druids see Piggott (1968) pp. 146–50. The excerpt from Blake's millenarian vision is from *Milton* I, 6, ll. 18–25, see Blake (1927). The correspondence between Borlase and William Stukeley is in the Morrab Library, see Pool (1986) pp. 125–30 for discussion.

## 15. MADRON

For the description of Cornwall, 'Cornwall! The Delectable Duchy! The Fairies Playground! The Land of Junket and Cream!...', see Dexter (1932) p. 1.

'The celebration of Christmas was formally abolished as a pagan remnant by an ordinance of 1647,' Walsham (2011) p. 134. 'The Holy Thorn at Glastonbury was subjected to further disfigurement by Roundhead forces, who cut it down in an act of "pure devotion", as John Taylor the water poet commented sarcastically,' ibid. p. 134. Jesus's encounter at

Jacob's Well is from John 4:20–21. For God's abolition of the 'distinction of places', see Walsham (2011) p. 82. George Joye's rage at those 'who runne after straunge goddes, into hilles, wodes and solitary places...' is quoted in ibid. p. 85.

For Abkhazian belief see Rachel Clogg's chapter 'Religion' in Hewitt (1999) pp. 205–17; 'Each lineage has its own sacred place, or *a'nyxa*... These sacred places are natural locations, high up in the mountains, or in forest-groves, by springs or rivers, cliffs or sacred trees,' ibid. p. 211.

The story of John Trelille comes originally from *The Invisible World* (1852) by the Calvinist and satirist Joseph Hall, and is quoted in Quiller-Couch (1894) pp. 127–9. For other instances of miracles and beliefs about Madron, see ibid. pp. 125–38.

## 16. ZENNOR

Borlase's theories about rock-basins are presented in Borlase (1973) pp. 240–58. His discussion of sacred places in antiquity, 'all places not equally auspicious', is from ibid. p. 122.

For W. G. Hoskins's citing of Zennor's field system, see Hoskins (1955) p. 28. An account of Frederick Hirst's life and work is in *Cornish Archaeology* 24 (Thomas, 1985b), and the manuscript of a fuller version of Thomas is in the Courtney Library, Thomas 4. Hirst's own archive is also in the Courtney Library.

More words, and more notions have been expended on the interpretation of stone circles than on any other element of prehistory. For a general overview of the tradition and those who have tilted at the stones with their theories see Hayman (1997) and Michell (1982). Horace Walpole, 'whoever has treated of [Stonehenge]...' is quoted in Leapman (2003) p. 199.

For information on the Oxford University expedition to Karahunj in Armenia, see Vardanyan (2010).

Francis Palgrave's observation, 'I was struck... by the likeness between the masses of rock, piled up by Nature only, and those cromlechs which also occur in Cornwall,' is recorded in Tennyson (1897) vol. I, p. 463. For the more recent idea of imitating the natural tors in the erection of quoits, see Tilley and Bennett (2001) pp. 335–62. Hollingham's intense experience of walking towards Land's End – 'We began to feel that we could walk like this for ever...' – is from Val Baker (1982) p. 34.

The ecstatic quote from Charles Dickens comes from Forster (1908) vol.1, pp. 285–6. Dickens also said of his journey to Cornwall: 'Such a trip as we had into Cornwall just after Longfellow went away!... Sometimes we travelled all night, sometimes all day, sometimes both... Heavens! If you could have seen the necks of bottles, distracting in their immense variety of shape, peering out of the carriage pockets!' The more sober reaction to the setting sun at Land's End was recounted by Forster himself, who was one of Dickens's companions on the trip (ibid. p. 285).

For *A Week at the Land's End*, see Blight (1861). The proof copy of Blight's *The Cromlechs of Cornwall* is in the archive of the Morrab Library, Penzance. 'I have a notion, of which I cannot be rid, that my brain is all wrong,' is from a letter he wrote to his friend and patron James Halliwell, quoted in Bates and Spurgin (2006) p. 135.

The announcement of Blight's 'death' by his publisher in 1884 – when he still had twenty-seven years to live – appeared in 'An Advertisement to the Second Edition of *Churches and Antiquities of West Cornwall*' (ibid. p. 223). It begins: 'By Mr Blight's death Archaeology has lost not only an enthusiastic fellow student, and it is much to be feared that this too eager devotion to his favourite pursuit amidst his daily toil brought on the illness that had so sad a termination.'

Until the late 1970s it was commonly thought that Blight had died in the 1870s, but the discovery of one of his asylum diaries in a house clearance helped establish the truth about his forty years at St Lawrence's in Bodmin. A second asylum diary surfaced in 2004 at a car-boot sale in Kent. Both these notebooks are now in the archive of the Morrab Library in Penzance. The picture of the outing to the Harlyn Bay Excavations in 1900 with Blight 'staring into the grave' is from the Royal Institution of Cornwall photo archive, ref. Harlyn Bay 38; see Bates and Spurgin (2006) p. 226.

## 17. NANJIZAL

On the origin of the name Costa da Morte, Carrera (1998) p. 18 cites the historian Henrique Rivadulla Porto's emphasis on ancient cosmology: 'Fisterra or Land's End in the ancient Greek and Celtic world stretched across the part of Galicia known as Dutika Mere, which stood for the region of misfortune or death, where the Helios, the sun, disappeared every day into the sea.' For the pre-Compostela pilgrimage to Finisterre in Galicia and the Galician folk story of the couple and their bed, see Romero (1993) p. 18.

George Borrow on his guide's terror of the *Estadea* is from Borrow (1923) pp. 422–3. For Richard Carew finishing his *Survey of Cornwall* with his shape-poem of Land's End, see Carew (1602) p. 160.

The various reactions of travellers on reaching Land's End are as follows: George Borrow in Fraser (1997) p. 32, Burritt (1865) p. 309, Craik (1884 ) pp. 112, 116.

W. H. Hudson (1908) recounts witnessing the pilgrimage to Land's End in his account of six months or so spent in the area of Land's End. He saw the pilgrims setting out daily from Penzance 'in batches of twenty-five or thirty or more', very varied in age and origin, but each intent on reaching the headland: 'the collection of unlike faces with the light of the same feeling in the eyes of all', ibid. pp. 38–9. The day – 24 May 1907 – when he saw the three train-loads of pilgrims arrive is from ibid. pp. 185–6. At the same time his wide reading of travelogues of Cornwall was 'an exceedingly wearisome task' lightened only by 'a kind of sporting interest' in what extravagant description the traveller might make of Land's End: 'the famous spot where he would have to pull himself together and launch himself bird-like from the cliffs, as it were, on the void sublime', ibid. p. 39. Hudson's own launching of himself produced his wonderful projection of the old men's vision of life beyond death, ibid. pp. 181–3.

## 18. LETHOWSOW

For Lethowsow/Lyonesse, see Carew (1602) p. 3, Hunt (1884) pp. 39–41, and Thomas (1985a) pp. 264–94. The haziness of the notion of Lyonesse is highlighted by what a visiting journalist said to Arthur Quiller-Couch: 'I had thought [Lyonesse] was just a poet's name for Cornwall,' in Quiller-Couch's Prologue to *Tre, Pol and Pen: The Cornish Annual 1928*, p. 6, co-edited by Charles Henderson in Henderson Collection, *Miscellaneous*.

For 'Sunk Lyonesse', see de la Mare (1922) p. 89. The Lyonesse quote from *Brideshead Revisited* is from Waugh (1945) p. 29.

Tennyson's embellishment of Lyonesse comes from a prose sketch of 1833, quoted in the memoir by his son, Tennyson (1897) vol. I, pp. 122–3. Tennyson at this time was already using the Arthurian stories as allegories for various aspects of Christian faith, and the places he described carried the same allegorical weight; see Burchell (1953) pp. 418–24. The lines from *Idylls of the King* are from Tennyson (1884), lines 260–64.

## 19. SCILLY

The *Elegy of Fortinbras* (trans. Czesław Miłosz) is in Herbert (1985) p. 98. John Fowles's reflection on islands is in Fowles (1978) p. 12.

For Solinus's and Sulpicius Severus's use of the singular in their reference to Scilly, see Thomas (1985a) p. 60. For archaeological work on Arthur, see B. St J. O'Neil, notebooks II, III, IV, archives of Scilly Museum, Hugh Town, St Mary's. O'Neil corrects Hencken's assessment of the ridge-top grave on Arthur in Hencken (1932) pp. 26, 318: it is not an entrance grave but a closed cist, O'Neil IV, p. 139. For Arthur as a 'pan-Brittonic figure of local wonderment', see Padel (1994) pp. 1–31. For an overview of work on Scilly's prehistory see Johns (2011) pp. 187–196; also see Robinson (2007) pp. 122, 134.

## EPILOGUE

The quote from *Peter Camenzind* is from Hesse (1953) p. 5. Regarding the siting of Cretan palaces, as above, see Scully (1962) p. 11.

# BIBLIOGRAPHY

The many and varied works that have fed into this book are too numerous to list. All books to some extent are the result of synthesis and I would like to acknowledge my debt to the great corps of academics and scholars who have helped map out the past as well as to those writers, novelists and poets who have discovered the surprising rewards to be had from exploring the 'spirit of place'. Like an elaborate cave system, it is a subject that gives little sign on the surface of the tunnels and immense chambers that lie beneath.

Below are only those books and articles referred to, or quoted from, in the text or in the notes.


Abrams, Lesley and Carley, James (eds), *The Archaeology and History of Glastonbury Abbey: Essays in Honour of the Ninetieth Birthday of C. A. Ralegh Radford* (1991) Woodbridge, Suffolk

Auden, W. H., Introduction to *Slick but not Streamlined* by John Betjeman (1947) New York

Balchin, W. G. V., *The Cornish Landscape* (1983) London

Baring-Gould, Sabine, *Adolescent Notebook Circa 1849–51: Mostly Filled at Pau, Bayonne, and Dartmoor*, transcribed by Ron Wawman (2010) www.nevercompletelysubmerged.co.uk/another-diary/adolescent-notebook10mb.pdf

— *Early Reminiscences 1834–1864* (1923) London

— *Mehalah* (1884) London

— *Cornwall: A Book of the West* with an Introduction by Charles Causley (1981) London

Barton, R. M., *A History of the Cornish China-Clay Industry* (1966) Truro

Bates, Selina and Spurgin, Keith, *The Dust of Heroes: The Life of Cornish Artist, Archaeologist and Writer John Thomas Blight 1835–1911* (2006) Truro

Bede, *Bede's Ecclesiastical History of the English People,* ed. and trans. Bertram Colgrave and R. A. B. Mynors (1969) Oxford
Bender, B., Hamilton, S., Tilley, C., *Stone Worlds: Narrative and Reflexivity in Landscape Archaeology* (2007) Left Coast Press, California
Blake, William, *Poetry and Prose of William Blake,* ed. Geoffrey Keynes (1927) London
Blight, John Thomas, *A Week at the Land's End* (1861) Penzance
Bond, Frederick Bligh, *The Gate of Remembrance: The Story of the Psychological Experiment Which Resulted in the Discovery of the Edgar Chapel* (1918) Oxford
Borlase, William, *Antiquities Historical and Monumental of the County of Cornwall* (revised edition originally published 1769, facsimile published 1973) Wakefield, Yorkshire
— *The Natural History of Cornwall* (1973, originally published 1758) London
Borrow, George, *The Bible in Spain: Or the Journeys, Adventures, and Imprisonments of an Englishman, in an Attempt to Circulate the Scriptures in the Peninsula* (1923) London
Boycott, A. and Wilson, L. J., 'In Further Pursuit of Rabbits: Accounts of Aveline's Hole, 1799 to 1921', *Proceedings of the University of Bristol Spelaeological Society* 2 (2011) Bristol
Bradley, Richard, *An Archaeology of Natural Places* (2000) London
— 'Ruined Buildings, Ruined Stones: Enclosures, Tombs and Natural Places in the Neolithic of South-West England', *World Archaeology* 30, no. 10 (1998) Oxford
Broadhurst, Paul and Miller, Hamish, *The Sun and the Serpent* (1989) Launceston
Brontës, The, *Tales of Glass Town, Angria, and Gondal: Selected Early Writings by the Brontës* ed. Christine Alexander (2010) Oxford
Burchell, S. C., 'Tennyson's "Allegory in the Distance"', *PMLA* (Journal of the Modern Language Association) vol. 68, no. 3 (1953) New York
Burritt, Elihu, *A Walk from London to Land's End and Back, with Notes by the Way* (1865) London
Burton, William, *Porcelain: Its Nature, Art and Manufacture* (1906) London
Carew, Richard, *Survey of Cornwall* (1602) London

Carley, James P., *The Chronicle of Glastonbury Abbey: An Edition, Translation and Study of John of Glastonbury's Cronica Sive Antiquitates Glastoniensis Ecclesie*, trans. David Townsend (1985a) London
— 'The Manuscript Remains of John Leland: "The King's Antiquary"', *Transactions of the Society for Textual Scholarship* 2 (1985b) Oxford
— 'John Leland in Paris: The Evidence of his Poetry', *Studies in Philology* 83 (1986) North Carolina
— *Glastonbury Abbey: The Holy House at the Head of the Moors Adventurous* (1988) London
— 'Leland, John (c.1503–1552)', *Oxford Dictionary of National Biography* (2006) Oxford
Carrera, Xan X. Fernandez, *Costa da Morte: Guía Turística-Cultural* (1998) A Coruña, Galicia
Casey, E., 'How to Get from Space to Place in a Fairly Short Stretch of Time' in Feld, S. and Baso, K. (eds), *Senses of Place* (1996)
— *The Fate of Place: A Philosophical History* (1998)
Causey, Andrew, *Peter Lanyon* (1991) London
— *Peter Lanyon: Modernism and the Land* (2006) London
Chope, Pease, *Early Tours in Devon and Cornwall* (1967) Newton Abbot
Clark, P. A., *The Neolithic Ritual Landscape of Rudston* (2004)
Clarke, Catherine, *Literary Landscapes and the Idea of England 700–1400* (2006) Cambridge
Clemo, Jack, *Confession of a Rebel* (1949) London
— *The Clay Verge* (1951) London
— *The Map of Clay* with an Introduction by Charles Causley (1961) London
— *The Marriage of a Rebel: A Mystical-erotic Quest* (1980) London
Collins, Wilkie, *Rambles beyond Railways or Notes in Cornwall Taken Afoot* (1851) London
Colloms, Brenda, 'Gould, Sabine Baring- (1834–1924)' in *Oxford Dictionary of National Biography* (on-line edition 2005) Oxford
Cox, Thomas, *Magna Britannia et Hibernia, Antiqua et Nova: Or a New Survey of Great Britain: Cornwall* (1720) London
Craik, Dinah Mulock, *An Unsentimental Journey through Cornwall* (1884) London
Cresswell, Tim, *Place: A Short Introduction* (2004) Oxford

Curtius, Ernst Robert, *European Literature and the Latin Middle Ages*, trans. William R. Trask (1953) London
Davidson, John, *The Last Testament of John Davidson* (1908) London
Davis, Simon R., *The Early Neolithic Tor Enclosures of Southwest Britain*, PhD thesis, University of Birmingham
de la Mare, Walter, *Down Adown Derry* (1922) New York
Dexter, T. G. F., *Cornwall: Land of the Gods* (1932) London
Dickinson, B. H. C., *Sabine Baring-Gould: Squarson, Writer and Folklorist 1834–1924* (1970) Newton Abbot
Du Maurier, Daphne, *Vanishing Cornwall* (1967) London
Dudley, Dorothy, 'The Medieval Village at Garrow Tor, Bodmin Moor', *Medieval Archaeology* (1962)
Eck, Diana, *India: A Sacred Geography* (2012) New York
Eliade, Mircea, *Patterns in Comparative Religion* (1963) Cleveland
Eliot, T. S., *After Strange Gods: A Primer of Modern Heresy*, Page-Barbour Lectures at the University of Virginia 1933 (1934) London
— *The Complete Poems and Plays of T. S. Eliot* (1969) London
Escobar, Arturo, 'Culture Sits in Places: Reflections on Globalism and Subaltern Strategies of Localization', *Political Geography* (2001)
Forster, John, *The Life of Charles Dickens* (1908) London
Fowles, John, *Islands* (1978) London
Fraser, Angus (ed.), *Penquite and Pentyre, or The Head of the Forest and the Headland: George Borrow in Cornwall 1853–1854* (1997) George Borrow Society
Friel, Brian, *Translations* (1982) London
Garlake, Margaret, *Peter Lanyon* (1998) London
Geoffrey of Monmouth, *The History of the Kings of Britain (Historia Regum Britanniae)*, trans. Sebastian Evans, revised by Charles W. Dunn (1958) London
Gildas, *The Ruin of Britain*, ed. and trans. Michael Winterbottom (1978) London
Giles, J. A. (ed.), *William of Malmesbury's Chronicle: Kings of England, from the Earliest Period to the Reign of King Stephen* (1847) London
Gilpin, William, *Observations on the Western Parts of England, Relative Chiefly to Picturesque Beauty* (1808) London

Gleeson, Janet, *The Arcanum: The Extraordinary True Story of the Invention of European Porcelain* (1998) London
Green, Thomas, 'A Gazetteer of Arthurian Onomastic and Topographic Folklore' in *Arthurian Notes & Queries* (1999, revised 2009) www.arthuriana.co.uk/notes&queries/N&Q2_ArthFolk.pdf
Harrison, George, *Memoir of William Cookworthy, Formerly of Plymouth, Devonshire by his Grandson* (1854) London
Hayman, Richard, *Riddles in Stone: Myths, Archaeology and the Ancient Britons* (1997) London
Heath-Stubbs, John, *A Charm against the Toothache* (1954) London
Heidegger, Martin, 'Building Dwelling Thinking' (1954) in *Poetry, Language, Thought*, trans. Albert Hofstadter (1975) New York
Heissig, Walter, *The Religions of Mongolia* (1980) London
Hencken, H. O'Neill, *The Archaeology of Cornwall and Scilly* (1932) London
Henderson, Charles, *Essays in Cornish History* (1935) Oxford
Herbert, Zbigniew, *Selected Poems* (1985) Manchester
Herodotus, *The Histories*, trans. Aubrey de Selincourt (1954) London
Herring, Peter, *An Exercise in Landscape History: Pre-Norman and Mediaeval Brown Willy and Bodmin Moor, Cornwall* MPhil (1986) Sheffield
— 'The Cornish Landscape' in *Cornish Archaeology* 50 (2011)
Hesse, Hermann, *Peter Camenzind*, trans. W. J. Strachan (1953) London
Hewitt, George (ed.), *The Abkhazians* (1999) London
Hoskins, W. G., *The Making of the English Landscape* (1955) London
Hudson, Kenneth, *The History of English China Clays: Fifty Years of Pioneering and Growth* (undated) Newton Abbot
Hudson, W. H., *Land's End: a Naturalist's Impressions in West Cornwall* (1908) London
Hunt, Robert, 'The Legend of the Lionesse', in *The Western Antiquary* vol. IV, 3 (1884) Plymouth
Hussey, Christopher, *The Picturesque: Studies in a Point of View* (second edition 1967) London
Innes, Hammond, *Wreckers Must Breathe* (1959) London
Jackson, Kenneth, *Studies in Early Celtic Nature Poetry* (1935) Cambridge

Johns, Charles, 'Ancient Scilly: The Last Twenty-five Years', *Cornish Archaeology* 50 (2011) Truro

Johnson, Francis, 'Two Treatises by Thomas Digges' in *The Review of English Studies*, New Series, vol. 9, no. 34 (1958) Oxford

Johnson, Nicholas and Rose, Peter, *Bodmin Moor: An Archaeological Survey, Volume 1: The Human Landscape to c. 1800* (1994) London

Jones, Andy; Taylor, Sean; Sturgess, Jo, 'A Beaker Structure and Other Discoveries along the Sennen to Porthcurno South West Water Pipeline' in *Cornish Archaeology* 51 (2012) Truro

Kendrick, T. D., *British Antiquity* (1950) London

Knox, Father Ronald and Leslie, Shane, *The Miracles of King Henry VI: Being an Account and Translation of Twenty-three Miracles from the Manuscript in the British Museum* (1923) Cambridge

Lanyon, Andrew, *Peter Lanyon 1918–1964* (1990) Newlyn

— *Peter Lanyon: Works and Words* (1995) Penzance

— *Wartime Abstracts: The Paintings of Peter Lanyon* (1996) St Ives

— *Peter Lanyon: The Cuttings* (2006a) Newlyn

— *Circular Walks around Rowley Hall* (2006b) London

Leapman, Michael, *The Troubled Life of Inigo Jones, Architect of the English Renaissance* (2003) London

Leopold, Aldo, *A Sand County Almanac, and Sketches Here and There* (1949) New York

Mabey, Richard, *Selected Writings 1974–99* (1999) London

Mackenna, F. Severne, *Cookworthy's Plymouth and Bristol Porcelain* (1946) Leigh-on-Sea

Malinowski, Bronislaw, *Malinowski and the Work of Myth*, ed. Ivan Strenski (1992) Princeton

Maltwood, Katherine, *A Guide to Glastonbury's Temple of the Stars: Their Giant Effigies Described from Air Views, Maps, and from 'The High History of the Holy Grail'* (1934) London

Mandelstam, Osip, *Journey to Armenia*, trans. Clarence Brown (1980) London

Marsden, Barry, *The Early Barrow Diggers* (1974) Aylesbury, Bucks

Michell, John, *Megalithomania: Artists, Antiquarians and Archaeologists at the Old Stone Monuments* (1982) London

Millard, Anne and Noon, Steve, *A Street through Time: A 12,000-year Journey along the Same Street* (1998) London
Mills, L. H., *The Zend Avesta, Part III, The Yasna, Visparad, Afrinagan, Gahs and Miscellaneous Fragments* (1887), Sacred Books of the East Series, vol. 31, Oxford
Milton, John, *The Poetical Works of John Milton* (1900) Oxford
Nicolson, Adam, *Sea Room: Life on One Man's Scottish Islands* (2001) London
North, Christine, 'Figs, Fustians and Frankincense: Jacobean Shop Inventories for Cornwall', *Journal of the Royal Institution of Cornwall* New Series (1995) Truro
Ogilby, John, *The South-West Highway Atlas for 1675* (2005) Ilkley
Ollard, Richard, *A Man of Contradictions: A Life of A. L. Rowse* (1999) London
Padel, Oliver, 'Cornish Place-name Elements', *English Place-name Society* vol. LVI/LVII (1985) Nottingham
— *A Popular Dictionary of Cornish Place-names* (1988) Penzance
— 'The Nature of Arthur', *Cambrian Medieval Celtic Studies* 27 (1994) Aberystwyth
Pausanias, *Description of Greece*, trans. J. Frazer (1898) London
Paynter, William, 'Daniel Gumb: The Cornish Cave-man Mathematician', *Old Cornwall* (1931–6)
Payton, Philip, *Cornwall: A History* (2004) Fowey, Cornwall
Penaluna, W., *An Historical Survey of the County of Cornwall in Two Volumes* (1838) Helston
Penderill-Church, John, *William Cookworthy, 1705–1780: A Study of the Pioneer of True Porcelain Manufacture in England* (1972) Truro
Piggott, Stuart, *The Druids* (1968) London
— *Ruins in a Landscape: Essays in Antiquarianism* (1976) Edinburgh
Polsue, J., *Lake's Parochial History of the County of Cornwall* 4 vols (1974) Wakefield
Polwhele, Richard, *Biographical Sketches in Cornwall* 3 vols (1831) Truro
Pool, P. A. S., *William Borlase* (1986) Truro
Pounds, N. J. G., 'The Discovery of China Clay', *Economic History Review*, New Series, vol. 1, no. 1 (1948) London

Pryce, William, *Mineralogia Cornubiensis: A Treatise on Minerals, Mines and Mining* (1778, facsimile edition 1972) Truro
Pryor, Francis, *Seahenge: A Quest for Life and Death in Bronze Age Britain* (2008) London
— *The Making of the British Landscape: How We Have Transformed the Land, from Prehistory to Today* (2010) London
Purcell, W. C., *Onward Christian Soldier: A Life of Sabine Baring-Gould, Parson, Squire, Novelist, Antiquary 1834–1924* (1957) London
Quiller-Couch, M. and L., *Ancient and Holy Wells of Cornwall* (1894) London
Rahtz, Philip, 'Pagan and Christian by the Severn Sea' in *The Archaeology and History of Glastonbury Abbey*, Lesley Abrams and James P. Carley (eds.) (1991) Woodbridge, Suffolk
Ratcliffe, Jeanette, *Fal Estuary Historic Audit* (1997) Cornwall Archaeological Unit, Truro
Relph, Edward, *Place and Placelessness* (1976) London
Robinson, C., *The Prehistoric Island Landscape of Scilly*, British Archaeological Reports, Brit. Series 447 (2007) Oxford
Romero, Fernando Alonso, *O Camino de Fisterra* (1993) Madrid
Rowse, A. L., *Little Land of Cornwall* (1986) Gloucester
— *The Diaries of A. L. Rowse*, ed. Richard Ollard (2003) London
Schulting, R. J., '...Pursuing a Rabbit down a Hole: New Research on the Early Mesolithic Burial Cave of Aveline's Hole', *Proceedings of the University of Bristol Spelaeological Society* (2005) Bristol
Scully, Vincent, *The Earth, the Temple and the Gods: Greek Sacred Architecture* (1962) Yale
Selleck, Albert, *Cookworthy 1705–80, and his Circle* (1978) Plymouth
Smith, Reginald A., 'The Rillaton Gold Cup' in *British Museum Quarterly* (1936) vol. 11, no. 1
Steinbeck, John, *A Life in Letters* (2001) London
Stephens, Chris, *Peter Lanyon: At the Edge of Landscape* (2000) London
Stubbs, William (ed.), *Memorials of Saint Dunstan, Archbishop of Canterbury* (1874) London
Swedenborg, Emanuel, *Arcana Coelestia*, trans. John Clowes (2009) Pennsylvania
— *The Doctrine of Life* (1954) Swedenborg Society, London

Sweet, R. H., 'Whitaker, John (1735–1808)', *Oxford Dictionary of National Biography* (2004 online edition) Oxford
Tennyson, Alfred, *Idylls of the King* (1884) London
Tennyson, Hallam, *Alfred Lord Tennyson, a Memoir, by His Son* (1897) London
Thomas, Charles, *Exploration of a Drowned Landscape: Archaeology and History in the Isles of Scilly* (1985a) London
— 'The Fiftieth Anniversary of the West Cornwall Field Club' in *Cornish Archaeology* 24 (1985b) Truro
— 'The Context of Tintagel: A New Model for the Diffusion of Post-Roman Imports' in *Cornish Archaeology* 27 (1988) Truro
— (ed.) 'Tintagel Papers', *Cornish Studies 16*, Institute of Cornish Studies (1989) Redruth
Thoreau, Henry David, *Walden* (Penguin Classics edition 1986) London
Thorpe, Carl, 'The Time Team in Cornwall' in *Cornish Archaeology* 50 (2011) Truro
Tilley, Christopher, *A Phenomenology of Landscape: Places, Paths and Monuments* (1994) Oxford
Tilley, Christopher and Bennett, Wayne, 'An Archaeology of Supernatural Places: The Case of West Penwith', *Journal of the Royal Anthropological Institute* vol. 7, no. 2 (2001) London
— 'The Powers of Rocks: Topography and Monument Construction on Bodmin Moor', *World Archaeology* 28, no. 2 (1996)
Todd, Malcolm, *The South-west to AD 1000* (1987) London
Toulmin Smith, Lucy (ed.), *The Itinerary of John Leland in or about the Years 1535–1543 Parts I to III* (1907) London
Trigger, Bruce, *A History of Archaeological Thought* (1989) Cambridge
Tuan, Yi-fu, *Topophilia: A Study of Environmental Perception, Attitudes and Values* (1990) New York
Turner, T. H. and Parker, J. H., *Some Account of Domestic Architecture in England* (1859) Oxford
Val Baker, Denys, *The Timeless Land: The Creative Spirit in Cornwall* (1973)
— *The View from Land's End* (1982) London
Vardanyan, Mihran, *Stars & Stones 2010*, Oxford University Expedition to Qarahunge, Armenia, www.qarahunge.icosmos.co.uk

Vincent, Nicholas, 'Richard, First Earl of Cornwall and King of Germany (1209–1272)', *Oxford Dictionary of National Biography*, online edition (2008), Oxford

Walling, R. A. J., *George Borrow: The Man and His Work* (1909) London

Walsham, Alexandra, *The Reformation of the Landscape: Religion, Identity, and Memory in Early Modern Britain and Ireland* (2011) Oxford

Watkins, Alfred, *The Old Straight Track: Its Mounds, Beacons, Moats, Sites and Mark Stones* (1925) London

Waugh, Evelyn, *Brideshead Revisited* (1945) London

Wawman, Ron (transcribed and annotated), *Sabine Baring-Gould's Adolescent Notebook Circa 1849–51: Mostly Filled at Pau, Bayonne, and Dartmoor* (2010) www.nevercompletelysubmerged.co.uk/another-diary/adolescent-notebook10mb.pdf

Weatherhill, Craig, *Cornish Place-names and Language* (1998) Wilmslow, Cheshire

Whitaker, John, *The History of Manchester*, second edition, 4 vols (1773) London

— *The Ancient Cathedral of Cornwall* (1804) London

— 'The History of Ruan Lanihorne' (ed. H. L. Douch), *Journal of the Royal Institution of Cornwall* New Series vii (1974) Truro

Whitley, H., 'The Silting up of the Creeks of Falmouth Haven', *Journal of the Royal Institution of Cornwall* (1881) Truro

Winchester, Angus J., 'Cookworthy, William (1705–1780)', *Oxford Dictionary of National Biography* (online edition, 2004) Oxford

Wordsworth, William, *The Poetical Works of William Wordsworth*, ed. Thomas Hutchinson (1895) London

# ACKNOWLEDGEMENTS

Particular thanks are due to Angela Broome of the Courtney Library and all the staff of the Royal Cornwall Museum; to Annabelle Read, John Simmonds and the staff of the Morrab Library, Penzance; to the staff of the Cornish Studies Library, Redruth; the staff of the Public Record Office, Truro; Vivien Stals and Linda Moffatt of Cornwall Library Store, Threemilestone, Truro; members of the China Clay History Society, St. Stephen; the staff of the London Library and of the National Archive, Kew. I am grateful to Charles Thomas, pioneer of landscape archaeology, and to Pete Herring for his encouragement, expertise and company on moorland trips and kitchen-table discussions. To Gillon Aitken and Andrew Kidd, to Philip Gwyn Jones, and to Laura Barber, for her diligence and panache, and to Daphne Tagg.

I am indebted too in all sorts of different ways to: Narmandakh Baatarjee, Barbara Bender, Sarnia Butcher, Mirabel and Hugh Cecil, Rose Cecil, Jason Cowley, Mark Crees, Stuart Croft, David Edgerton, Christine Edwards, Roger Farnworth, Stephen Fitt, Charles Fox, Patrick Gale, Michael Galsworthy, Derek Goodwin, Robin Hanbury-Tenison, Richard Hoare, Anthony Hobson, Will Hobson, Miles Hoskin, Aidan Hicks, Ian Jack, Kurt Jackson, Paul Jackson, Tigran Khzmalyan, Graeme Kirkham, Andrew Lanyon, Tim Law, Jeremy Le Grice, Lyn Le Grice, Jude Le Grice, Rafael Lema, Keith Lowe, Barbara Luke, Sue and Christopher Marsden-Smedley, Amanda Martin, George Miller, Oliver Padel, Brian Perman, Luke Piper, Jane Powning, Manuel Rivas, Ann Soutter, Ben Tay, Ash Taylor, Mark Tetley, Caroline Tetley, D.M. Thomas, Carl Thorpe,

Julie Throssel, Nick Tomalin, Tony Toms, Jane Turnbull, Anna Tyacke, Chris Varcoe, Stephen Varcoe, Tom Varcoe, Sarah Vivian, Craig Weatherhill, Jason Whittaker, Mark Willson, Linda Wilson.

And a loving thank you to Clio and Arthur who help to prevent it all getting too serious, and to Charlotte, of course, for judicious assessing of ideas, for tireless critical reading, and for all manner of less visible support, as ever.

# ILLUSTRATION CREDITS

On Garrow Tor (p. 43) and Rough Tor (p. 61) by Richard Hoare. Tintagel, 1813 (p. 81) by Letitia Byrne, and Charles Henderson (p. 185) with thanks to the Courtney Library. *Glastonbury Tor* (p. 99) by Luke Piper. Higher Ninestones (p. 117), and Jack Clemo (p. 135) with thanks to the China Clay History Society, St Stephen. The will of Robert Bennett of Tregony, 1607 (p. 151) with thanks to the Cornwall Record Office. *Porthleven (Ship)* by Peter Lanyon (p. 205) © Sheila Lanyon all rights reserved DACs. *Cornwall* (p. 225) from *Plot for Sale* by Andrew Lanyon. Madron Baptistery (p. 239) from Cyrus Redding, *An Illustrated Itinerary of the County of Cornwall*, 1842 p. 172. Rude Stone-Monuments (p. 251) from William Borlase, *Antiquities of Cornwall*, 1769, p. 158.

All other photographs by the author.

# INDEX